The Event Thread
for
Requirements

Bill Wiley

THE EVENT THREAD
FOR REQUIREMENTS

iUniverse books may be ordered through booksellers or by contacting:

iUniverse
1663 Liberty Drive
Bloomington, IN 47403
www.iuniverse.com
1-800-Authors (1-800-288-4677)

Because of the dynamic nature of the Internet, any web addresses or links contained in this book may have changed since publication and may no longer be valid. The views expressed in this work are solely those of the author and do not necessarily reflect the views of the publisher, and the publisher hereby disclaims any responsibility for them.

Any people depicted in stock imagery provided by Thinkstock are models, and such images are being used for illustrative purposes only. Certain stock imagery © Thinkstock.

ISBN: 978-1-4917-4726-1 (sc)
ISBN: 978-1-4917-4725-4 (e)

Library of Congress Control Number: 2014917450

Printed in the United States of America.

iUniverse rev. date: 12/22/2014

Table of Contents

Book 2
The Event Thread and Methodology

7.
Methodology Introduction

8.
Introduction to Use Cases

9.
Domain Model

10.
Business Event List

11.
High-Level Event Use Case

12.
Event Use Case JIT Detail

13.
The Event Advantage

14.
Physical Phases - Brief Discussion

Book 3
Case Study

Prologue

Since WWII, no super power has had a decisive victory over an insurgency; the record is 0-20 [Ramo, 2009, p. 89]. The super powers approach a conflict directly under a rigid model, a rigid set of plans. As a conflict takes on unexpected direction, the super powers just can't make adjustments easily or quickly enough [Ramo, 2009, p. 211]. The advantage the insurgency groups seem to have is their *ability to adapt, to remain agile.*

Development of information systems can suffer a fate similar to the super powers. As the business changes or as we find errors in the requirements, our systems can't be easily modified and the teams that build them just can't adjust quickly enough. The structure of the systems is too rigid and the development process can't adapt in a reasonable amount of time.

There is a *partitioning scheme* that can jumpstart the system development effort and extend through the lifecycle to deployment; it will affect virtually every aspect of the project. Natural partitions become common threads that tie the various lifecycle phases together from early to late in the lifecycle and retain a high recognition factor with both the business team and the IT team. *A system is produced whose components have low coupling and are resilient and will make the system and the teams both more adaptive and more agile.*

Introduction

With all of the advances in technology in the past couple of decades, large and small organizations alike are still having difficulty implementing successful information systems. Many projects are late and/or produce applications that are out of date or just not the correct solution.

For most large problems, divide and conquer is still the preferred approach. As a partitioning scheme, events define natural business activity and retain their identity throughout the development lifecycle. The resulting Event *Threads* extend naturally through the lifecycle and clarify the development process for both the IT and business teams.

The primary objective of this book is to build the business case for events and the resulting "Threads." *Herein are concise notes* on the discovery, definition, and documentation of system requirements and of the event partitions that emerge from a study of the user domain. As concise notes, it is not intended to provide detail explanations of each concept; the reader might need to research certain concepts depending on his/her experience level and background.

This book uses business events as the basis for defining system requirements. It is about how a proposed system can be defined and partitioned and the role these partitions can play in the definition of requirements. Events that occur naturally in the business and trigger a system response become the basis for the partitioning scheme and these partitions persist throughout the project lifecycle. But more important than the events are the Threads that result. These Threads extend from the early discussions to implementation; they transform the lifecycle.

The lifecycle phases along with change and risk management benefit from this natural, strong, pervasive partitioning scheme. The Thread can be applied to software packages and outsourced software development to assess and manage the match to requirements.

This is not an all-purpose project guide. Its focus is the set of events that drive and partition the system space. The early chapters in this guide are not necessarily in chronological order across the lifecycle and can be read in any order. One chapter is devoted to one or more topic/technique from the lifecycle diagram (Appendix A) and the relationship between the techniques is found in the diagram.

The Event Thread is composed of three books. Book 1 is ideal for the manager, project lead, and systems analyst who want to explore the concepts of the Event Thread. If she/he likes what they see, then Book 2 looks at the practical application of events. Book 3 presents a case study of six events that can be used to explore the application of the concepts of Books 1 and 2.

Book 1 is not about methodology. It presents the Event Thread and describes its value as a basis for methodology. *The most important aspect of this work is the underlying concept of business events and the partitioning of a proposed system into responses to those events. A common Thread for each event runs throughout the development process.*

Business events are intuitive to the user and are typically accepted by both the user community and the development group. They get the user group involved early in the development lifecycle by defining, in the user's language, relevant natural activities that occur in the business area. They also keep the user engaged throughout the development lifecycle. Business events also help reduce the communication gap that often arises during the software development effort; most every team member will recognize an event such as Customer Places Order.

In addition, the events partition the proposed system into subsystems that have relatively low coupling and support incremental development and implementation. These partitions are both pervasive and persistent; the Thread of a single event/system response extends through the project lifecycle from beginning to end and touches nearly every aspect of the project.

Chapters 2 and 3 discuss business events and the partner system responses that are fundamental to the *event-driven approach*. Chapter 4 presents the application of events to the development of *Use Cases*.

> *I have chosen Use Cases for this book because they dovetail nicely with events and quite frankly, I like a simplified version of them. However, there are many other ways to document requirements and the method chosen does not impact the effectiveness of events as the partitioning scheme. So for the remainder of this book, insert your favorite method of documentation when you see the words "Use Case."*

Chapter 5 introduces the middle-out concept in contrast to a top-down approach. Chapter 6 presents the key concept of the book, that is, the common *Event Thread* that runs the length of the lifecycle for each event.

The Thread is an important concept, but it must have a practical side to be of value to an IT team. Book 2 presents how to put The Thread to work. The early conceptual phase produces a high-level overview of the proposed system. The later conceptual phase creates sufficient detail for the system build.

Book 3 is a case study. It allows the reader to study the results of a methodology that partitions early and persists throughout the lifecycle. It will help deepen the understanding of the underlying partitioning scheme. *The running case study* is a simple merchandize order system such as might be found on the Web. In addition to order-related transactions, it allows for the return of an item, management control of discount information, and generation of various reports.

The three books will likely be used at times for reference purposes and often read independently. Consequently, by design, the reader will find some repetition throughout the book.

This guide is written for IT professionals and some experience in the definition of system requirements is expected. Use Case experience is not needed because the Use Case concept is not difficult and there are numerous sources that describe their use in detail. A Use Case is just one way to document requirements for one event response.

The reader can take away the following from this book.

- an understanding of the nature of an event and the resulting Thread;
- the strong connection between events and the user;
- the strength of the partitioning scheme as each Thread has its own lifecycle and artifacts;
- how a Thread preserves the identity and traceability of the event throughout the lifecycle;
- the application of the middle-out strategy to the definition of requirements;
- the natural relationship between events and Use Cases;
- a hybrid methodology using the Event Thread and Use Cases for an agile, adaptive approach to building an information system;
- an understanding of how an agile approach to building software differs from the traditional waterfall approach;
- based on the case study, examples of the event-driven concepts and methods at work.

How do *JAVA action events* fit into these concepts? Event threads are common to the JAVA world. A physical implementation of JAVA action events is similar in concept to the use of events for the definition of requirements described in this book.

A JAVA action event occurs when an action is performed by the user. Examples are when the user clicks a button, chooses a menu item, or presses Enter in a text field. The result is that a message is sent to all action listeners that are registered on the relevant component.

As an example, an application uses five radio buttons to let the user choose which of five special characters is displayed. There is a radio button for each special character. If the *caret* button is selected, the *caret* character is displayed. If a *tilde* button is then selected, the *caret* becomes unselected and the image switches from a *caret* to a *tilde*. Each time a character is selected, the button fires an action event and the application responds by deselecting the current choice and selecting and displaying the new character.

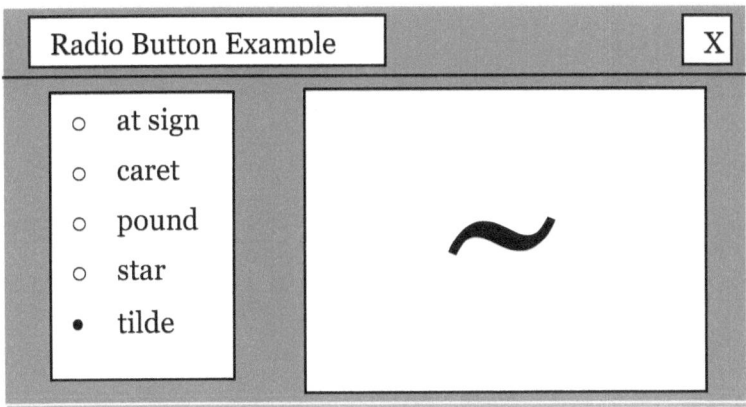

This example follows the structure of a requirements conceptual event, that is, a trigger and system response with stored data interaction as needed. However, this JAVA event is a physical implementation. This book is about the conceptual (not physical) definition of system requirements and *in no way is associated with a JAVA event.* It will not address the physical JAVA events again. The reader must keep focused on the conceptual.

Book 1
The Event Thread

Book 1 provides foundational information about event-driven and just-in-time requirements definition. Business events are born in the business space and represent natural business activities. They partition the system and are the basis for subsequent definition of each system response as an *Event Thread* extends from beginning to end of the project lifecycle.

Chapter 1
Foundational Concepts

This chapter introduces some of the key concepts that are necessary for an understanding of the primary components of the event approach (business event, event trigger, system response, and the Event Thread) and of the process for Use Case definition.

The Event Thread

The **Event Thread** is the key concept of this book. An Event Thread originates in the user space as an event and is a common thread that runs through most components of the project lifecycle. It partitions and defines scope for a Use Case and allows the development teams to focus on a relatively small, independent subsystem; it guides development and project management. Not one component of the project lifecycle escapes untouched by the Event Thread; an event is traceable back to the event list from any system component. It carries with it a high recognition factor across all teams (for example: *Customer Places Order* is an event that extends across the entire lifecycle and eventually is deployed as *Fill Customer Order*). This book will refer to this conceptual persistence as the Event Thread, the event partition as it extends down the project lifecycle from the event list to implementation.

Events: Natural Business Activity

Events occur in the system's environment. They happen in the user's world and are the natural business activities that users experience on a day-to-day basis; they are described in the business language of the user. Both conceptually and operationally, relevant events (events that fall within the scope of the proposed system) will invoke a response from the system. An example of an event is *Customer Places Order*.

The Event Trigger

When an event that is within the system scope occurs, some part of the proposed system is notified and is designed to respond. This notification is called the event trigger. *There is one and only one trigger for each event.*

Event Trigger. An incoming data flow that originates from an event and signals the system to take action, that is, to execute the planned system response. Also the stimulus that signals an occurrence of the initiating event.

Examples from the cyber order system are 'order' which originates from the event *Customer Places Order* and 'returned item' from the event *Customer Returns Item.*

The System Response

The system response is a chunk of the system designed to respond when its particular event occurs. It is highly cohesive and is loosely coupled to other partitions of the system. It is a planned response to fulfill the requirements defined for its particular event and is a natural system building block. A system response interfaces with the user and the stored data and not with other system responses.

> *Coupling*. The degree to which one system component is dependent upon another.
>
> *Cohesion*. The degree to which a system component accomplishes one and only one function.

Partitioning

Partitioning represents the breaking of a proposed system into pieces (partitions). The more effective partitions have high cohesion and low coupling. With an event approach, a planned system response to an event becomes one partition.

Events and Middle Out

Work done at the University of California at Berkeley by Eleanor Rosch has demonstrated that humans typically don't categorize and classify from the top down [DeSmedt, 1994, p.64]. They instead begin at a level somewhere in the middle with those things with which they are most familiar.

When a system is partitioned using events, something very similar is accomplished. The beginning is somewhere in the middle of the hierarchy (with familiar user activities) and the model is synthesized upward and decomposed downward as needed, a process referred to as "middle out." It is not a top-down approach in which the system is viewed as a single entity and then successively decomposed.

Events and the User

Events become a common thread between the business team (the users) and the IT team (the builders) and they typically define much of the graphical user interface.

Events and Use Cases

Events are a natural forerunner to a Use Case. Each event/system response has a corresponding Use Case (a one-to-one relationship). The event/system response sets the scope for the Use Case.

Events and Data

The early business language along with the middle out nature of events gives the Data Analyst an early look at the major data entities of the system. The state of a data entity changes in response to an event. For example, when the event *Order is Shipped* occurs in an online ordering system, the state of the order changes from "filled" to "shipped." In the data model fragment that represents the customer entity and the order entity, these events affect the relationship between the two data entities. They also begin to reveal attributes (for example, *order status*).

```
Events:  Customer places order
         Customer order accepted
         Customer order filled
         Customer order shipped
```

The early event list contains information about both data entities and the relationships that will be documented in the data model (see Appendix F for a normalized data model). These data requirements can be investigated on an event-by-event basis.

Managing Change

Changes can be isolated to a partition with a business identity. Because an event extends throughout the development lifecycle, changes can be traced from the business space to the corresponding requirement, design, and code. An event is, by its very nature, some activity in the business and a change to the business will likely be reflected in an associated information subsystem (system response). Also, due to the independence of event-partitioned responses, a new event response can be added, often without affecting existing parts of the system.

Managing Risk

As with change, risk assessment can be isolated to a partition with a business identity. Because an event extends throughout the development lifecycle, risk can be examined and traced from the business space to the corresponding requirement, design, and code.

Capturing and Verifying Requirements

The documentation and verification of requirements is key to the successful completion of an information system project. The system must not only be error free but more important satisfy the needs of the user community and the organization. The earlier the requirements are verified, the less expensive repair becomes when errors or omissions are discovered.

With events and the resulting Threads, requirements capture and verification can be focused on a single Thread, partitioning the effort and simplifying management of the process. Since a Thread has its own lifecycle and artifacts, each can be treated as a subsystem in its own right.

As software packages and the outsourcing of software proliferate, a match to the user needs must be confirmed early and the final product verified and tested. The thorough definition of requirements coupled with a strong, natural partitioning scheme is vital to confirming the correct system is being purchased.

The Process

The methodology represented in this book is a combination of waterfall and iterative phases in the project lifecycle. The early conceptual phase follows a waterfall approach and is a relatively quick pass through the proposed system creating the high-level artifacts shown below.

- o Domain Model
- o Event List

- System Response List
- Event Use Case Diagram
- High-Level Use Cases
- Acceptance Plan
- Deployment Plan

The later conceptual phase is iterative and addresses one Use Case at a time, passing each Use Case detail to subsequent design, build, and test efforts. Once successfully tested by the user (UAT), the component is available for release based on the deployment plan.

The entire process is presented in diagram form in Appendix A.

Chapter 2
Events: Natural Business Activity

*An event is a natural business activity
that is identified early in the lifecycle
and that persists to deployment
becoming a natural partition of the
system. An event is the root of one
common Thread that will stretch the
complete length of the lifecycle.*

The concept of business events was formally introduced by

McMenamin and Palmer in their 1984 book [McMenamin, 1984].

Event. "An event is some change in the system's
environment" [McMenamin, 1984].

Business Event. An activity in the user's environment
that requires a response from the proposed
information system [author]. This term and 'event' will
be used interchangeably in this book.

Screen event. An activity such as the clicking of a
command button in a screen window that invokes
some action from the software. Not related to a
business event and will not be considered further in
this book.

A business event and an event response (a.k.a. 'system response')

as defined for this book are taken from "Essential Systems

Analysis" by Stephen McMenamin and John Palmer: "An *event* is

some change in the system's environment, and a *response* is the set of actions performed by the system whenever a certain event occurs" [McMenamin, 1984]. "When a system's response to an event has been determined before the event occurs, then the interactive system generates a *planned response*. "Planned response systems don't react to every event in the environment. Many external and temporal events don't even raise a yawn from a particular system" [McMenamin, 1984].

A Natural Approach

Events are a natural approach to requirements definition because they occur in the business space and *become the natural building blocks of the proposed system*. They are the natural business activities that users experience on a day-to-day basis. They are described in the business language of the user and the user participates significantly in the definition of the events. Each event identifies the business role that will initiate contact with the system, the action taking place in the business area, and an abstract representation of the trigger that will invoke the system response.

The Nature of Events

The events that are of interest in this book are of two types: 1) external events, which are initiated by entities in the environment and 2) temporal events, which are initiated by the passing of time [McMenamin, 1984] or by the occurrence of a specified time.

For an *external business event*, some change in the system's environment occurs and sends an event trigger to the system, that is, the system is somehow notified that the event has occurred. (Event triggers will be discussed in a subsequent chapter as part of the system response). The part of the system dedicated to responding to the business event is in an idle state. It reacts to an instance of the trigger by completing the required processing and producing the required output. It then returns to an idle state. It is an elementary process and must leave the business in a consistent state [McMenamin, 1984.] By definition, all processing in response to a business event is continual (this does not mean that the system cannot pause) and must be accomplished before the event response becomes idle awaiting another instance of its trigger.

For a *temporal business event*, a point in time is reached. The part of the system dedicated to responding to that business event executes and completes all of the processing required and returns to an idle state. As with an external event, it must leave the business in a consistent state.

A third type of event, called a *state event*, is typically found in real-time systems. They invoke processing based on a change in system state, such as in an elevator control system or an automated painting system. This type of system is also a 'planned response' system (as described by McMenamin and Palmer) and can be modeled in a manner similar to external and temporal events. However, state events will not be discussed further in this book as they are not part of the typical business system.

By the definition of events and event responses, there are no internal events that trigger additional system processing. Any function that can respond immediately to the processing of an event response is, by definition, part of that response. Any function that responds to the processing of an event response in a delayed fashion is responding to a trigger from one of the two types of events and is an event response in its own right. A system response interfaces with its environment or the database and not other responses. *System responses communicate between themselves by leaving the database in an updated state.*

While business events occur outside of the proposed system (that is, in the business space), they indirectly partition the system. That is, the system responses to the events divide the system into highly independent 'chunks'. This is a key concept of this book and will be explored in depth in later chapters.

In-Scope Events

The events establish the scope of the system in terms of the business area activities that will be involved with the system, that is, to which the system must respond. While events are happening continuously in a business environment, only a subset is of relevance to the proposed system. Many business events get no reaction from the system. For example, the arrival of a customer at a retail store gets no reaction from an inventory system but the system will react if the customer subsequently buys something.

Each relevant event has an associated system response, that is, sixty business events will have sixty associated system responses. Relevant events will always invoke a response from the system. As this book uses the term business event, it typically will be referring to those events within the proposed system scope. Examples of relevant business events for an order processing system are *Customer Requests Order Status* and *Management Submits Item Discount Data.*

The Process

With the beginning of requirements definition, work is begun in the business arena. Natural business tasks are identified that will require a system response and thus be in the system scope. This phase of system development takes place on the business turf and is driven by the business team. In most cases, the IT team will lead and document this definition effort with the business team providing the business activity expertise.

During requirements definition, a list is developed documenting the business area events. These events are both pervasive and persistent as they affect most components of the system and last throughout the lifecycle. Each business event (external and temporal) and the resulting planned response begin to partition the proposed system into manageable pieces very early in the requirements definition phase. Subsequently, a Use Case will be developed for each event/system response pair.

The system partitions derived from the business events persist from the requirements phase through construction, testing, and deployment. They can be managed and developed independently based on data dependencies and priorities.

The role of the event is depicted below.

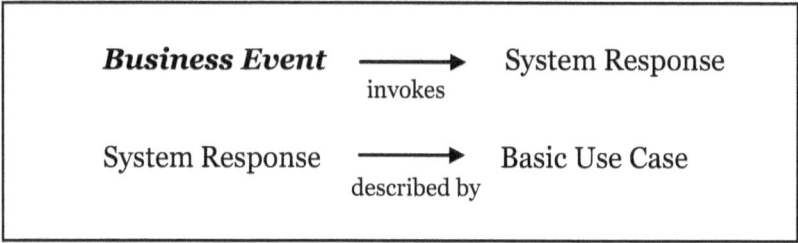

See Book 2 for the application of the Event Thread in the form of a methodology.

Chapter 3
The System Response

While a natural business event occurs outside of the proposed system (in the business space), the response to an event is part of the system and partitions the system space. It extends the Event Thread into the proposed system becoming a natural system building block.

The System Response is the processing that must be completed by the proposed system when an instance of a business event occurs.

System Response. The set of actions performed by the system whenever a certain event occurs [McMenamin, 1984].

System Response. The processing that must be completed when an instance of the corresponding business event occurs [author].

This allows the early definition of the proposed system in terms of high-level partitions. Something sufficient to build software certainly has not been defined but we now have a building block that will eventually become part of the system. Just how many of these (one or 100) begins to reveal the size of the future effort.

Each business event identified in the early requirements definition effort shall have a corresponding system response; there is a one-to-one relationship. If there are 60 business events, there will be 60 system responses. Each system response becomes a 'chunk' or partition of the proposed system and is represented in the abstract so that a physical solution is not implied. The entire system is composed of these partitions.

The system response has one and only one mission in life, to complete the required processing when an instance of the corresponding event occurs. The system response remains in an idle state until the event occurs, completes all of its assigned processing, and then returns to an idle state awaiting another instance of its partner event. A system response interfaces with its business environment and the database and not other system responses directly. *It communicates by leaving the database in an updated state.*

An example of an event/response pair is *Management Submits Item Discount Data/Apply Discount Data*. The system response is triggered by arrival of discount data (the trigger) and responds by applying that data to the discount data store and confirming the update.

Trigger. A signal to the system from outside the system that a specific event has occurred. It is represented by a data flow into the system originating from the occurrence of an event.

Planned Response System

Systems that respond to business events are referred to by McMenamin and Palmer as 'planned response systems'. An agent of the system's environment acts on the system, and the system reacts to the external agent. Whenever a specific business event occurs, a related set of actions is performed by the information system in response to the event trigger. Conceptually a trigger is a data flow into the system originating from the event's user interface or in the case of a temporal event, the passing of time. The relationship of the system response to the business event and Use Case is shown below.

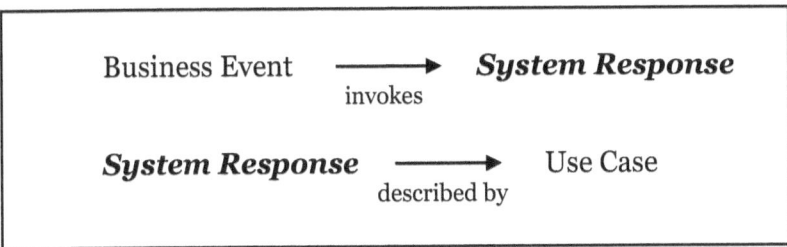

Chapter 4
Events and Use Cases

A Use Case that is based on a business event has a foundation that provides the user interface, scope, and partitioning properties. An Event Thread that began in the business space as an event and emerged in the system space as a planned system response must now be defined as a requirement.

Once the business events and the associated system responses have been identified and have partitioned the proposed system, the requirements for each system response must be specified. For the approach in this book, a basic Use Case will be developed for each business event/system response pair. A Use Case defines the system response. It documents the interaction between the system user and the system and describes the necessary work to be completed, including interaction with the database.

There is more than one way to build a Use Case. And there are many books and Web sites that provide detailed instruction for building them. The goal of this chapter is to describe how business events and the system response drive the definition of Use Cases (and not to instruct how to build a Use Case).

> *Basic Use Case.* Describes a sequence of actions, between the actor (the user) and the system, which is invoked by an external trigger and provides something of value for the actor.

A Use Case can be used to capture functional requirements and is a black box with no internal system structure defined. *What* the system will do is described, not *how* it will be done (design). A Use Case represents one chunk of the system and its scope is determined by the need to fulfill a single user request (event). A Use Case defines the user interface and describes what the system must accomplish to generate the required output and to achieve the Use Case goal. The required data elements are also documented. A typical system will have dozens of events and thus dozens of Use Cases.

> Use Case name examples:
>   *Return Item Availability*
>   *Fill Customer Order*
>   *Generate Weekly Sales Report*

Event-Driven Use Case

An event-driven Use Case is one that is derived from a *business event* (previously defined in the Event List). The event provides the trigger that invokes the Use Case processing. Conceptually there is a one-to-one mapping between a business event and a basic Use Case. However, a physical implementation of reuse might generate Use Cases that are not mapped directly to a business event.

Business Event ——————▶ System Response
　　　　　　　　　　invokes

System Response ——————▶ ***Basic Use Case***
　　　　　　　　　described by

A Use Case will be developed to specify each event/system response in the proposed system. The advantage of defining business events first is that they occur in the business space and are close to the users' natural everyday activity. They are described in the language of the business.

Once the business events and the system responses are identified, development of the Use Cases can begin. In the waterfall phase, each Use Case is very high level. In the iterative phase, each Use Case is developed fully, just-in-time for the design and build.

Chapter 5
Partitioning

A partitioning strategy that is based on natural business activity and persists throughout the lifecycle provides a strong foundation for agility in system development and maintenance. An Event Thread is born with each partition.

System Partitioning

The objective of system partitioning is to divide both the conceptual and physical systems into 'chunks'. Partitioning of the proposed system space is a major outcome of the early conceptual phase. It is pervasive; it affects most aspects of the subsequent system development process. The strategy chosen for this effort has consequences that persist throughout the development lifecycle. Partitioning allows the development teams to work on smaller, cohesive pieces of the proposed system. The partitions simplify project, change, and risk management tasks since each partition is a virtual subsystem to be managed separately.

The partitioning scheme should produce independent subsystems that will support subsequent lifecycle efforts. Early methodologies as well as some more recent ones decompose from the top down. These partitioning schemes can result in system components that have high coupling, that is, significant dependence upon each other. This dependency increases the development and

41

management effort throughout the development lifecycle as well as for post-deployment maintenance and it greatly increases the impact of change.

Event Partitioning

When a system is partitioned using business events, the beginning is somewhere in the middle of the hierarchy with familiar user activities (events). This is not a top-down approach in which the system is viewed as a single entity and then successively decomposed until an adequate level of detail is obtained. The use of business events as the partitioning scheme is a central theme of this book.

> *To partition by events.* To divide the system into major components based on the system's response to natural business events. Each system response to an event becomes a system partition.

Using events and the associated system responses as the primary scheme for partitioning the proposed system brings with it many advantages, from the way it influences the entire development lifecycle to the way in which it links the system to the natural business activities. Partitioning begins early with the identification of events in the business space and extends through the system lifecycle. These natural business partitions jumpstart involvement by the user in the definition of requirements. They are described in the users' language and they focus on those users' work activities that relate to the system under study.

A response to each business event is defined by its user interface, its processing requirements, and its interface to stored data and is encapsulated in a single conceptual unit. In effect, the event-driven user interface is extended to influence the partitioning of the entire system.

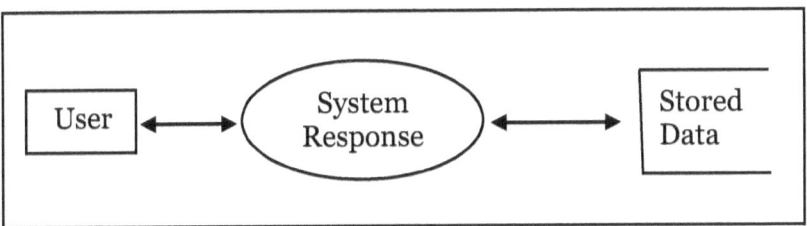

One of the benefits of event partitioning is that the event responses are coupled only through the stored data making them highly independent. The data dependencies are documented in an interaction model and they influence the priority of subsystem development and deployment as the system transitions from conceptual requirements to a physical configuration.

Interaction Model. Models the interaction between the data and the system responses. Examples are: CRUDA Matrix, Entity Process View, Entity Lifecycle Diagram [Wiley, 2000].

As development moves through the physical tasks, the event response subsystems remain independent. The design team works with the business events that were identified by the user in the earliest tasks of the project and still remain intact. This brings familiarity, consistency, and control to the project. And since they also continue to be recognizable as business activity throughout the development process, they facilitate ongoing communication between the business team and the development team.

Events simplify many project management tasks by allowing the allocation of resources to relatively simple, highly cohesive subsystems and allow for flexibility in the assignment of scope priorities. They also aid in other project management activities since each partition is a subsystem to be managed separately.

Both risk management and change management can be applied to each event partition, simplifying these ongoing tasks. The extension of the events throughout the lifecycle aids traceability from test results to code back to requirements.

Chapter 6
The Event Thread

*Each business event identified in the
early stages of the project becomes the
origin of a common Thread that
extends through the lifecycle bringing
a single focus to each lifecycle task and
artifact.*

The business event (a.k.a. event), the system response (a.k.a.
event response), and the Use Case are three key elements of the
conceptual phase approach presented in this book. Their common
thread, a Thread that stretches from beginning to end of the
lifecycle, is the focus of this chapter.

Natural business events occur in the business space. Once a
business event is determined to be within the proposed system
scope, it requires a partner system response. These system
responses, in total, define the system and divide the proposed
system space into many partitions. But what occurs subsequently
is even more powerful. Each event and system response pair
retains its identity throughout the lifecycle; this becomes the
common Thread that brings the lifecycle phases and tasks to a
single focus and ties the development artifacts together for each
partition. *Each **Event Thread** has its own lifecycle.* The
following diagram represents an Event Thread cutting through
the lifecycle layers.

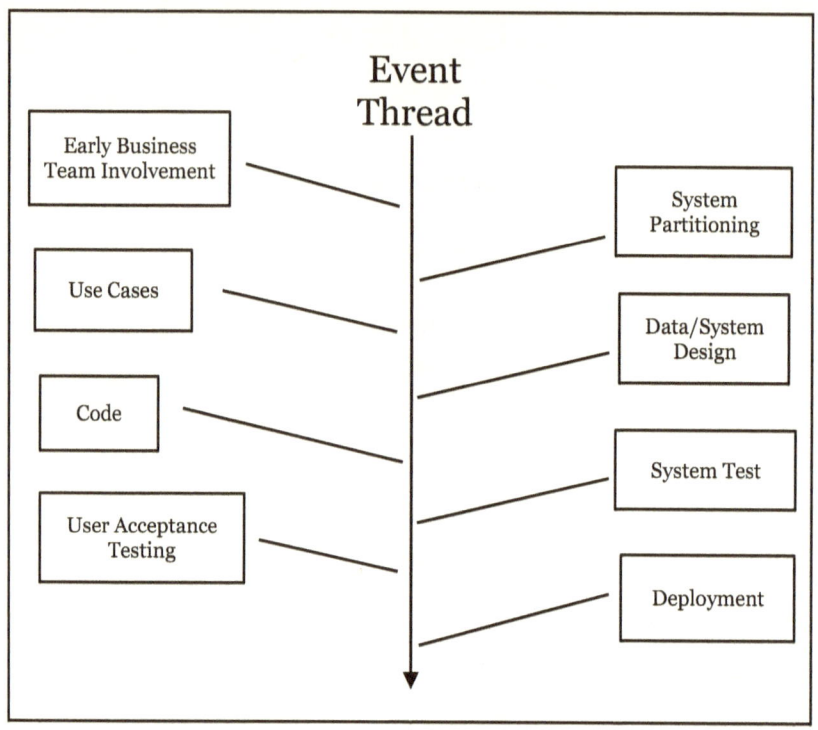

For example, the event *Customer Places Order* and its associated system response *Fill Customer Order* establish a Thread that passes through requirements, design, build, test, UAT, and deployment; this event can be treated as a separate subsystem. *Fill Customer Order* can be found in each lifecycle phase and artifact and extends into the production environment. *The Event Thread concept alone could transform the outcome of many projects!*

> *Event Thread.* The common element that runs
> through each lifecycle phase and artifact based
> on the event that lay at its origin in the natural
> business activity. Each Thread has its own
> lifecycle. A proposed system will have dozens of
> events and Event Threads.

Events Extended

Finding business activities (events) can be thought of as a
horizontal (two dimensional) effort that typically identifies
dozens of events. But each event identified in the initial discovery
JAD sessions establishes a Thread that **extends vertically**
through the lifecycle from requirements definition to the
deployment of system components. Because an Event Thread has
its own lifecycle, it slices through the entire development process
providing a single event perspective throughout.
Diagrammatically the Thread for *Customer Places Order* is
viewed in the third dimension as it cuts through the various
lifecycle layers.

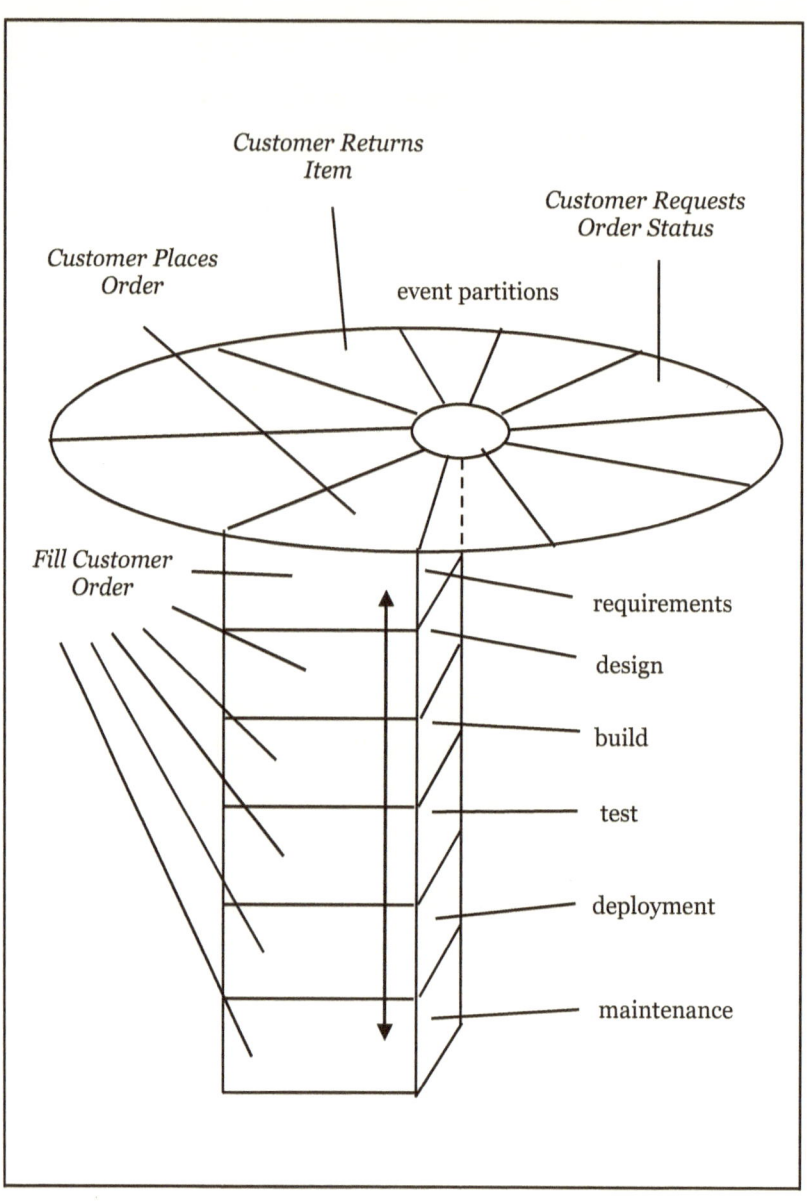

Customer Returns Item

Customer Requests Order Status

Customer Places Order

event partitions

Fill Customer Order

requirements

design

build

test

deployment

maintenance

Natural Business Activity

Each event and thus each Event Thread have at their origin a **natural business activity**. This establishes a familiarity with each system partition that aids in communication between and within all teams involved for the life of the project. As the project focus shifts across the lifecycle, the business team remains in touch as each Event Thread retains the identity of the business activity that lay at its root.

Recognition

Another key benefit is the **recognition** factor of each event as it extends through the lifecycle. Virtually every member of the project teams will recognize the Event Thread for *Fill Customer Order* regardless of the lifecycle phase or artifact. The Thread is recognizable during requirements definition to the later tasks such as User Acceptance Testing (UAT) and deployment and can typically be found in the system's graphical user interface. Not one component of the project lifecycle goes untouched by the Event Thread! An illustration of the recognition factor and traceability for *Fill Customer Order* follows.

- Domain Model: *customer, incoming order*
- Event List: *Customer Places Order*
- System Response Table: *Fill Customer Order*
- Lean Use Case: *Fill Customer Order*
- Acceptance Test Plan: *Fill Customer Order*
- Deployment Plan: *Fill Customer Order*
- JIT Use Case: *Fill Customer Order*

- Results Visibility: *Fill Customer Order*
- Data Requirements: *Customer* and *Order*
- Design, Build, Test: *Fill Customer Order*
- User Acceptance Test: *Fill Customer Order*
- Deployment: *Fill Customer Order*
- GUI: *Customer Places Order*
- Maintenance: *Fill Customer Order*

Cohesion

The partitioning and the Event Threads that evolve allow the teams to **focus** on a relatively small, highly cohesive chunk of the system one at a time. This focus comes into play at the earliest sessions of the project when events are being identified and provides benefits from requirements definition all the way to the build and deploy efforts. The supplementary tasks of project, change, and risk management also benefit from being able to focus resources and ongoing management attention on the loosely coupled system partitions.

Traceability

At times, an issue arises in a component of one phase of the lifecycle that requires examination of or change to the related component(s) of the event. Because of the pervasive nature of the Thread, **tracing** backward or forward to an associated component is easy and natural. If a design issue arises for *Fill Customer Order*, it is easy to identify the Use Case holding the requirements (*Fill Customer Order*) and the event and the associated subject matter expert (SME). Or if an acceptance test for *Fill Customer Order* fails, the preceding components such as functional requirements or data requirements can be easily traced to the physical code modules.

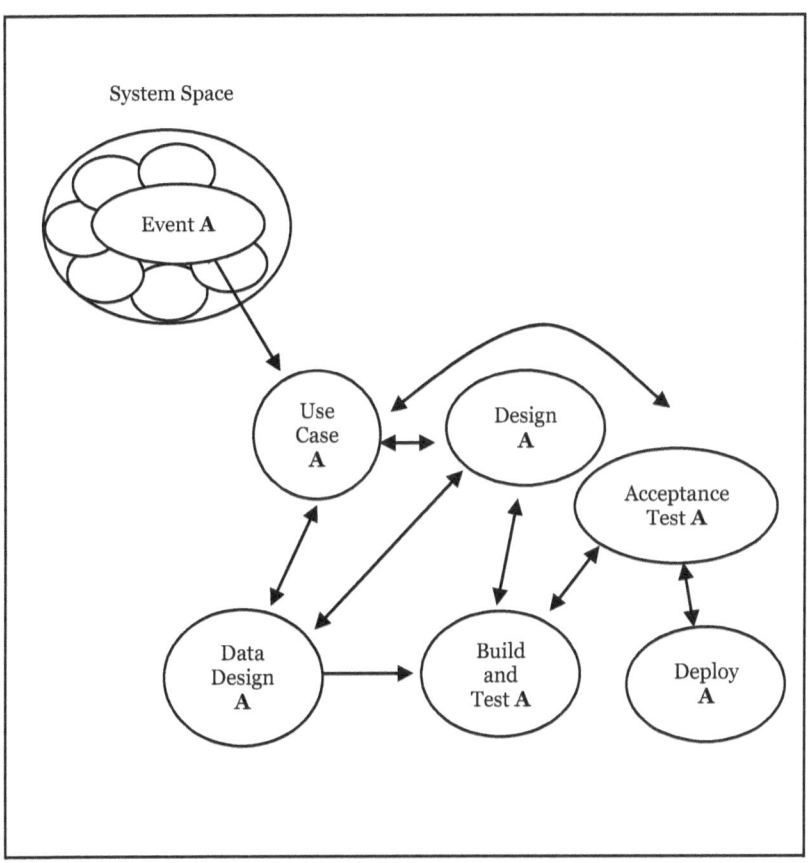

Verification

Possibly the most important of the benefits of the Event Thread comes during verification of the requirements. Unit, integration, and system level testing do not verify the correctness of system requirements. These tests just verify that the requirements, correct or incorrect, were developed as specified and that the software runs without errors. The only tests designed to verify requirements are the user acceptance tests (UAT). These tests are based on the requirements and check to see that the correct solution from a business perspective was developed. The problem with discovering errors and omissions at this point in the development lifecycle is that UAT occurs after development and is much too late. When possible, errors need to be discovered during the conceptual modeling phase when the cost of repair is the lowest.

With an event strategy and the associated Threads that occur, **verification** of the requirements can focus on a Use Case and on the associated user acceptance test that will ultimately be used for final confirmation of the correct build. A SME and the appropriate requirements analysts can isolate an effort to carefully examine the contents of a single Event Use Case and to confirm or repair the requirements. Changes at this point are many times less costly to make than those detected later in the lifecycle.

Agility

One of the main themes running through Joshua Ramo's "The Age of the Unthinkable" is the need for government, business, and individuals alike to adopt an agile life strategy, one that has the ability to adapt built in. *There is a lesson here for both business and IT*. The world order has changed and the 'unthinkable' must be expected in nearly all aspects of life; a less stable world is upon us.

"We are now tied to one another in ways we can't see, through webs of finance or disease or information, and – here's the dangerous paradox – the more closely we're bound, the less resilient we all become." [Ramo, 2009, p. 198] As the world order changes, as the world becomes more unstable and less predictable, *so will the business climate.*

As Ramo points out, insurgencies use their *adaptive nature* to an advantage when facing major world powers, resulting in a startling record (insurgencies 20, world powers 0) [Ramo, 2009, p.88-89]. Major Powers on the other hand put everything they have into a direct, rigid, frontal attack with little flexibility.

IT often does something similar. They amass everyone for requirements, then design, build, and test across months or even years. But this approach has failed over and over. What is needed is a more agile approach where making adjustments is part of the plan, where adapting is the norm. The ability to adapt during the development process and into the production phase is vital. The development teams should seek out needed changes in the system as development progresses.

Typically, requirements begin to age soon after they are defined. Defining detailed requirements for an entire proposed system then spending the next year building that system only assures that an out-of-date system will be delivered. To build from requirements that are months old increases the odds of a system that does not meet business needs. Defining and verifying requirements just-in-time allows delivery of a current system. In an event strategy, change has manageable impact because relatively small, isolated Event Threads are dealt with. A change is less risky to implement because it is less likely to have a cascading effect that causes scope and cost and the impact of change to escalate.

IT can take the lead in preparing for change when designing and building new information systems. The teams must be comfortable updating the system model, and in fact should expect and plan to do so.

Adaptive system. One to which new requirements can be added and existing ones dropped or changed with minimal impact.

While agility depends a lot on the skill set and the mindset of the development teams, event partitioning and the resulting Threads provide a platform and system structure that is foundational for an adaptive strategy.

Deployment

Another advantage of the Thread can be found in the system deployment strategy. Threads can be developed in parallel and a subset of the total system deployed for user checkout or, less likely, for production purposes. The development strategy can be based largely on the Use Case priority but deployment must also take into consideration the population of the database as data must first be written to the database before it can be accessed. *Often database population is the primary barrier to deploying a system subset for early production.*

Book 1 Epilogue

The concept of system partitioning is not new. However, many of the early schemes were more complex than the problem they were meant to solve. Partitioning with events produces highly cohesive "chunks" that have very low coupling. These partitions persist and are pervasive and retain their identity throughout the lifecycle.

Book 1 has provided foundational information about the Event Thread. Business events occur naturally in the business space and partition the system. The resulting Threads extend from beginning to end of the lifecycle dividing the development effort into many independent strands that retain identity and traceability across the project. They are the basis for the two-phased, Use-Case driven methodology discussed in Book 2.

Book 2
The Event Thread and Methodology

Theory is good. But if that theory fails to bring advantage to the application phases, one could question its real value.

Book 2 explores the application of the Thread. The early conceptual waterfall phase takes a one-pass, high-level look at the proposed system. Events and lean Use Cases are the product of a relatively quick assessment of the system to be built. It serves to orient the teams and establish high-level scope. Possibly the most important outcome of this early phase is to partition the system into loosely coupled event subsystems.

The late conceptual phase is an iterative, just-in-time process that focuses on a single Use Case. Fresh requirements are passed downstream to the design, build, test, and deploy steps one Use Case at a time.

The Event *Thread* continues to bind and unify the various artifacts and lifecycle layers from the earliest methodology steps to deployment.

Chapter 7
Methodology Introduction

*The theory and concepts presented in
Book 1 form a foundation for a
practical application in a methodology
setting. The Thread affects
methodology from the earliest tasks to
deployment.*

The conceptual components described in this chapter are part of a
two-phase methodology: an early waterfall phase and a late
iterative, just-in-time (JIT) phase. These components will be used
to demonstrate the pervasiveness and persistence of the *Event
Thread* and how it will behave in the real world of a methodology.
It will also demonstrate the importance of the role of events
throughout the lifecycle. Refer to the "Event Thread" diagram in
Appendix B.

This book does not contain a detailed representation of the
methodology; it is used only to demonstrate event behavior. For
this discussion, refer to the diagram in Appendix A.

Waterfall Phase

A relatively quick-pass early waterfall phase brings with it a
number of advantages. It serves to bring the business team and
the IT team together early in the project in activities that are high
level and serve to orient all involved to the system development
task at hand. It jumpstarts the project.

Three of the early process components are driven by the events: domain model, event list, and high-level Event Use Cases. The acceptance plan and the deployment plan are also influenced heavily by the events. The artifacts produced by this phase are:

- o Domain Model (Context Diagram, Lean Data Model)
- o Event List
- o System Response Table
- o Use Case Diagram
- o High-Level Event Use Cases
- o Acceptance Test Plan
- o Deployment Plan
- o Assessment of Preliminary Timelines

Iterative Phase

The late conceptual phase is iterative and produces just-in-time (JIT) detail requirements for one Use Case per iteration. These requirements are defined just before being sent downstream to the build steps so are as fresh as possible. Each iteration also defines the results visibility (reporting) along with data requirements then verifies the contents of each Event Use Case. The artifacts produced by this phase are:

- o Detail JIT Event Use Cases
- o Interaction Models
- o Results Visibility (reporting)
- o Event Use Case Acceptance Tests
- o Data Requirements

The Case Study

The *running case study* is a simple merchandise order system such as might be found on the Web. In addition to order-related transactions, it allows for the return of an item, management control of discount information, and generation of various reports. Six of the system events are shown below (many other events exist for this system).

1. Customer Places Order
2. Customer Requests Item Availability
3. Customer Requests Order Status
4. Time to Generate Weekly Sales Report
5. Management Submits Item Discount Data
6. Customer Returns Item

Case Study Business Rules

A system is ultimately driven by the business rules. Following are nine rules that will determine system direction for this case study.

1. invalid credit card will be reported within 1 hour
2. order will not be filled if order amount takes credit card over limit
3. backorder will be shipped within 5 business days
4. orders out of stock will be offered substitute products
5. orders will not be delivered to post office box
6. only customers on file can access item availability
7. after two failed tries, session will be dropped
8. discount data is effective for a specified range of days
9. customer data must be verified before order is allowed

Chapter 8
Introduction to Use Cases

*A Use Case defines the user/system
interface and identifies the system
processing steps needed to produce the
required output. A Use Case defines the
requirements for a single Event
Thread.*

Using Use Cases for requirements capture is part of a user-centered analysis and they are the foundation for user acceptance testing. Use Cases describe the system from the user's point of view.

A Use Case represents a chunk of the proposed system. Events partition the system and a Use Case defines the functional requirements of an event response and identifies the user interfaces. A Use Case describes the dialogue between the system and a user (an actor in UML) necessary to achieve a particular goal for the user.

> A *Use Case* defines the interactions between an external
> user and the proposed system to accomplish a specific goal.

A Use Case is a description of a system's behavior, represented as a sequence of steps, as it responds to a request that originates from outside of that system. It describes the interaction between one or more users and the system itself, detailing scenario-driven paths through the functional requirements.

An agent specifies a role played by a person or thing when interacting with the system such as a customer or the Sales Department or another system (for example: an inventory system). Agents are something or someone which exists outside the system under study, and that take part in a dialogue with the system to achieve some goal. The same person using the system may be represented as different agents because they are playing different roles. Conceptually, a Use Case can communicate only with agents outside the system (not with other Use Cases).

The detail of a Use Case emerges as the project progresses. A Use Case can be brief at the beginning then become more detailed as more is learned in the later stages of the lifecycle.

Interactions with the system are perceived as from outside the system. What the system must accomplish will be the focus and not how it will be accomplished. This reduces the chances of making implementation decisions too early in the lifecycle and thus influencing and limiting the system requirements.

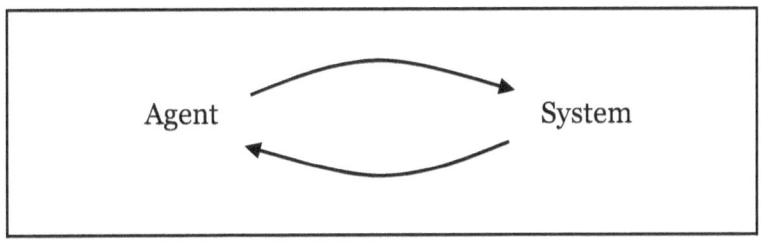

Scenario

Scenarios provide an important perspective when developing and validating a Use Case. A user of a Use Case can follow many paths as they interact with the Use Case. A Use Case has a primary path through the requirements and can also have one or more alternate paths. Each path is an instance of the Use Case, a scenario; each scenario is one complete path through the Use Case. Subsequently, a user acceptance test case will be developed for each scenario.

Each of the paths shown below is a scenario. The diagram does not represent all paths through the *Fill Customer Order* Use Case, for example, invalid product number and rejected payment are not shown.

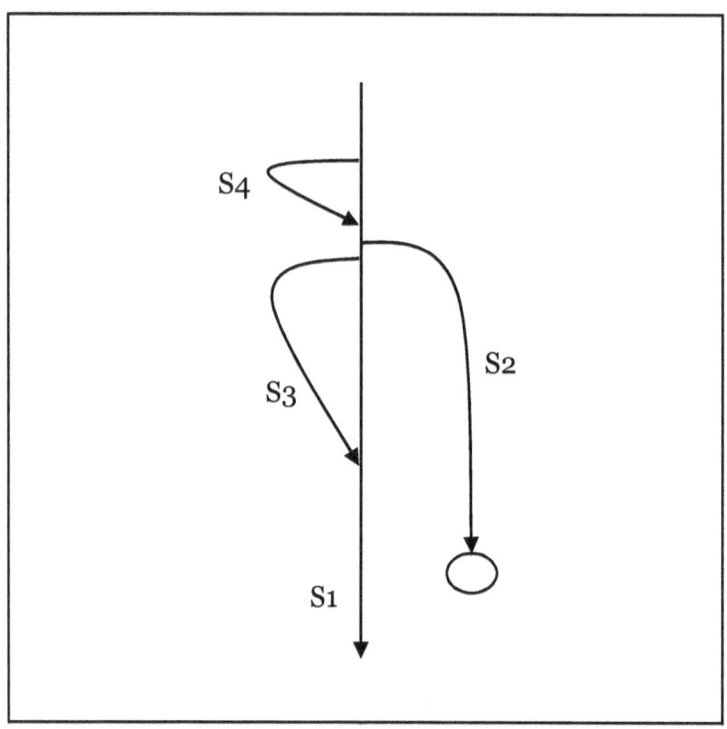

Use Case: *Fill Customer Order*

S1: successful order

1. customer selects 'place order' from menu
2. customer submits order information
3. system validates order information
4. system checks for sufficient quantity of product
5. system requests pay information from customer
6. customer submits pay information
7. system exports pay data
8. system updates data stores
9. system generates pick list, invoice, shipping documents

10. system informs customer of successful completion of order
11. customer closes order session

S2: rejected backorder

1. leaves S1 following step 4
2. system informs customer of insufficient product quantity
3. system offers customer product backorder choice
4. customer rejects product backorder option
5. system offers option to order another product
6. customer ends order session

S3: accepted backorder

1. leaves S1 following step 4
2. system informs customer of insufficient product quantity
3. system offers customer item backorder choice
4. customer accepts backorder option
5. return to S1 Step 5

S4: invalid order

1. leaves Scenario 1 following step 3
2. system reports invalid order
3. system offers corrective action
4. customer submits corrected order
5. return to Scenario 1 step 3

Use Case Scope

A scenario is invoked by an instance of the partner event and runs until the scenario goal is complete. The processing can pause but conceptually runs continually until the scenario goal is met then becomes idle awaiting another instance of the event. A Use Case must contain all possible paths.

Use Case Template

There is no standard template for documenting detailed use cases and individuals are encouraged to use templates that work for them or the project they are on. Standardization within each project is more important than the detail of a specific template. There is, however, considerable agreement about the core sections; beneath differing terminologies and orderings there is an underlying similarity between most use cases [Wikipedia: FSTda627600, *Use Case Templates*]. A Use Case has all or a subset of the following elements (components). Refer to Chapters 16 - 21 for examples of completed Use Cases.

1. Use Case Name: An active verb phrase (verb-noun format) that reflects the Use Case goal (example: *Fill Customer Order*).

2. Use Case Identifier: reference number, author, creation date, modification history, parent system

3. Use Case Priority: how critical this Use Case is to your system / organization in relationship to other Use Cases (will be used to make development and implementation decisions).

4. Goal: A description of the goal to be achieved by the Use Case; what the user expects to achieve when the Use Case runs.

5. Use Case Summary: A brief textual description made up of the Use Case name, user interface, and description of work to be completed including major inputs and outputs. This can be taken from the Lean Use Case created in the first pass of an iterative Use Case development methodology.

6. Use Case Trigger: The action upon the system (business event) that starts the Use Case; may be a time (temporal) event.

7. Actors: The entity that initiates this Use Case and all users who participate in this Use Case.

8. Preconditions: Activities that must be completed, conditions that must be met, states that must exist for the Use Case to begin.

9. Primary Scenario Steps: A detailed description of the successful interactions between the actors and the system (user actions and system responses) which are necessary to achieve the Use Case goal (from trigger to goal delivery).

10. Alternative Paths: All successful and unsuccessful paths through the Use Case other than the primary scenario.

11. Post Conditions: The state of the data following successful completion of the Use Case.

12. Business rules: Those rules or policies, written or unwritten, that determine how an organization conducts its business. They must be relevant to this Use Case.

13. Input Summary: A list of the data that are required from users or data stores.

14. Output Summary: A list of the data output by the system to the actors or data stores.

15. Use Case Issues: Unresolved questions about the Use Case.

16. Use Case Notes: Information that is not directly part of this Use Case but that needs to be considered in other Use Cases and other aspects of the project.

17. Non-Functional Requirements: Requirements that affect the design and/or implementation of the Use Case but don't directly affect the interaction between the actor (user) and the system response. Examples are security and requirements which impose constraints on the design or implementation such as performance requirements, quality standards, or design constraints.

Use Case Diagram

A Use Case Diagram provides a visual representation of the Use Cases and the system interface to its actors. A Use Case diagram can be developed early to get an overview of the Use Cases and the actors that will interface with the system or it can be built in parallel with the lean Use Cases. The role of each actor and the complexity of the actor interface are revealed.

The system-level Use Case diagram often has value to the architecture team as it reveals function distributed across the system.

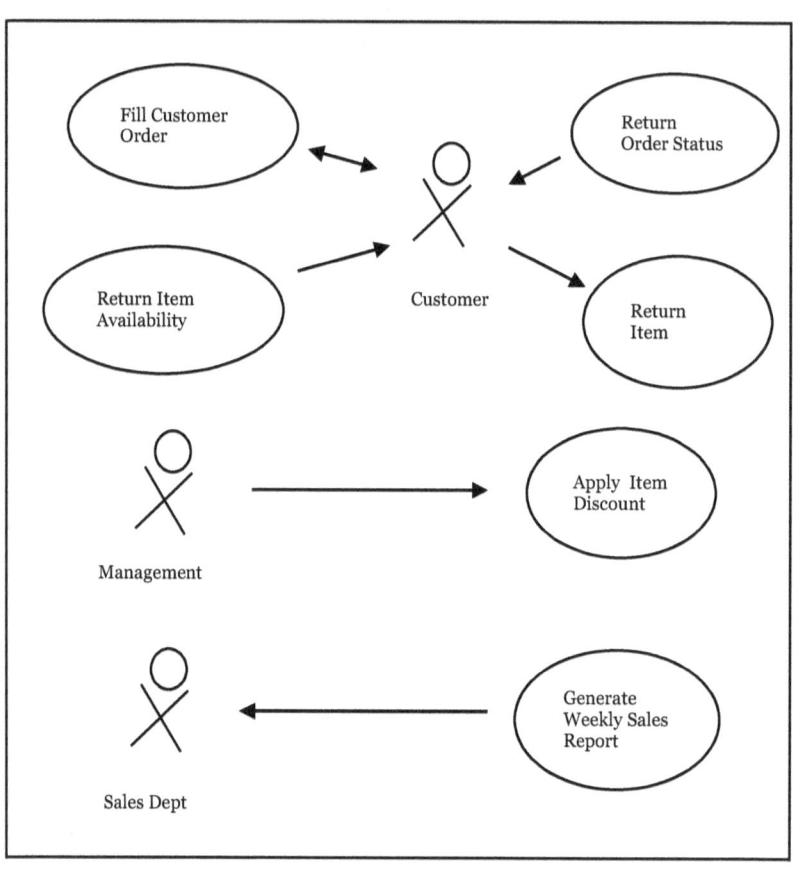

Fill Customer Order

Return Order Status

Return Item Availability

Customer

Return Item

Management

Apply Item Discount

Sales Dept

Generate Weekly Sales Report

Chapter 9
Domain Model

*The Domain Model defines the
system's business context. The relevant
business areas, the players, and the
major inputs and outputs are
identified. The Thread has its earliest
beginnings in this model as events
emerge from the input streams.*

Early in the project lifecycle, a general look at the relevant
business areas and major data entities involved in the proposed
system can *jumpstart the definition process* and pay big
dividends down the road. Domain modeling provides a starting
point for discussion between the business and IT teams preceding
requirements definition. It also helps identify who should be
present at subsequent joint requirements gathering sessions. And
the business events have roots in this model.

The Domain Model consists of the *Context Diagram and a Lean
Data Model* and documents high-level data flow, data entities,
and data relationships. This initial examination of the proposed
system in its context exposes user groups, major inputs, and
major outputs; it provides a high-level perspective throughout the
project. These diagrams will help the team begin to think about
scope and will help guide them during the identification and
definition of the functional requirements. Throughout the
lifecycle, the diagrams provide a reference for the system under

development and provide check points to ensure that all interfaces to the system are being considered.

Domain Modeling. Domain modeling consists of exploring and explaining the environment of the problem to be solved and its interfaces to the proposed system, and also identifying major entities. The resulting model provides an overall framework in which to add business events. A context diagram and high-level data model make up the domain model.

In the context example later in this chapter, interaction of the customer with the system reveals an incoming order, a returned item, a request for order status, and a request for item availability. *The event partitioning has begun as these data flows will spawn events in the subsequent event list.*

As entities and relationships are identified in the lean data model, more potential business events are revealed and along with the context diagram, *the set of events for the proposed system is born.*

The following depicts the beginning of the *Customer Places Order* event as an incoming order from a customer. Each subsequent chapter will track this Event Thread through the lifecycle.

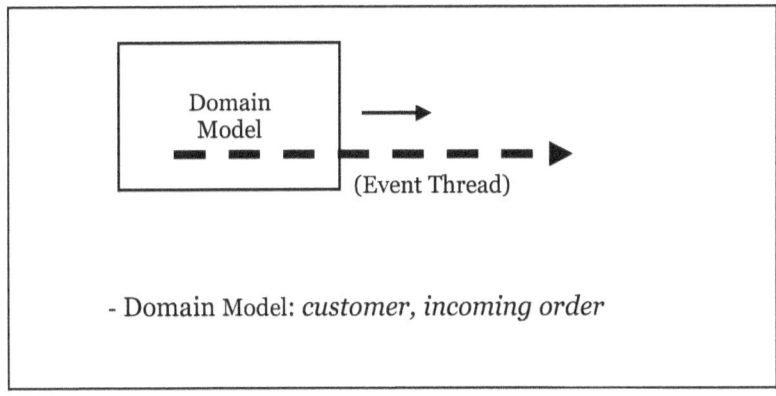

- Domain Model: *customer, incoming order*

Context Diagram

Procedure Overview

Meet with the business team to discuss the context of the proposed system and develop the context diagram. This diagram can be used to motivate and encourage dialogue early in the requirements definition effort and to capture a high-level view of the system interfaces.

o   The entire system is represented by a single DFD process.

o   Identify the major inputs to the proposed system that originate outside the system and to which the system must respond.

o   Identify external agents (providing input or receiving output) and the reasons for their interaction with the proposed system. Identify the input flow and output product for each agent.

o   When representing textually, use the normalized form (agent/data flow/output product).

o   Add external databases to which the system will interface (*Purchasing* in the Context Diagram that follows) since they might reveal and document a very complex system interface which needs to be understood early.

- o To keep the diagram readable, show composite data flows when the inputs and outputs begin to crowd the diagram.
- o After a first-cut diagram is complete, iterate through it as the events are discovered since system agents and their interface with the system will be revealed by the events.

From the case study

A partial list of domain paths in agent/data flow/output format:

- o customer/ order/ invoice (represents physical item ordered)
- o customer/ item (returned)/ return confirmation
- o Sales Department/ temporal/ weekly sales report
- o warehouse/ ship data (pick list, destination, schedule)
- o management/ discount data/ update confirmation

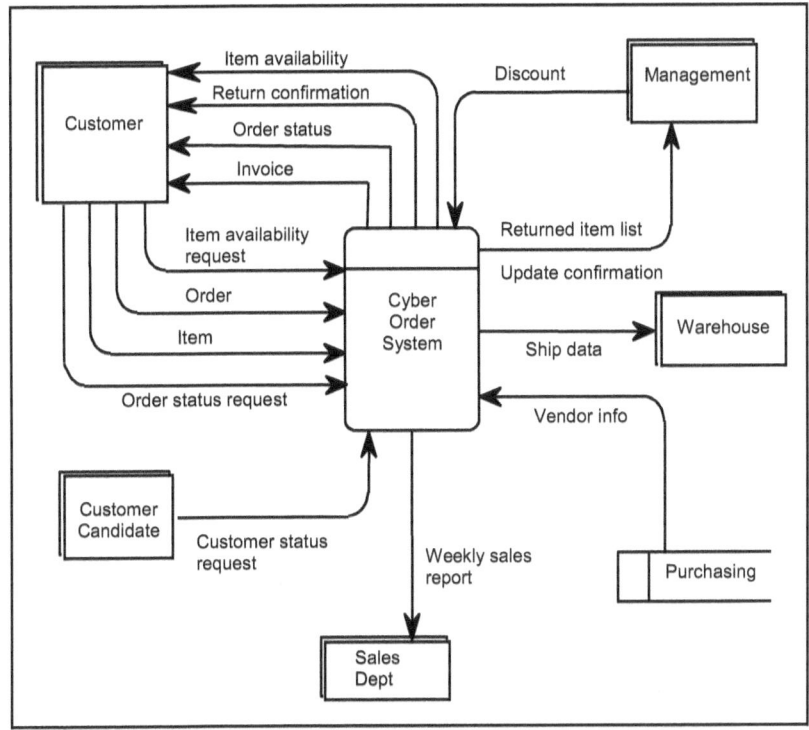

Events: events emerge as the customer inbound data flows are identified.

- customer places order
- customer returns item
- customer requests item availability
- customer requests order status

High-Level Data Model

Procedure Overview

Identify the data entities that are evident from the context discussion and diagram. Establish initial relationships based on the business rules known at this time.

From the case study

Partial initial data model (entity/ relationship/ entity):

- customer/ places/ one or many order
- order/ contains/ one or many item
- item/ contained on/ zero or many order
- order/ owned by/ one and only one customer

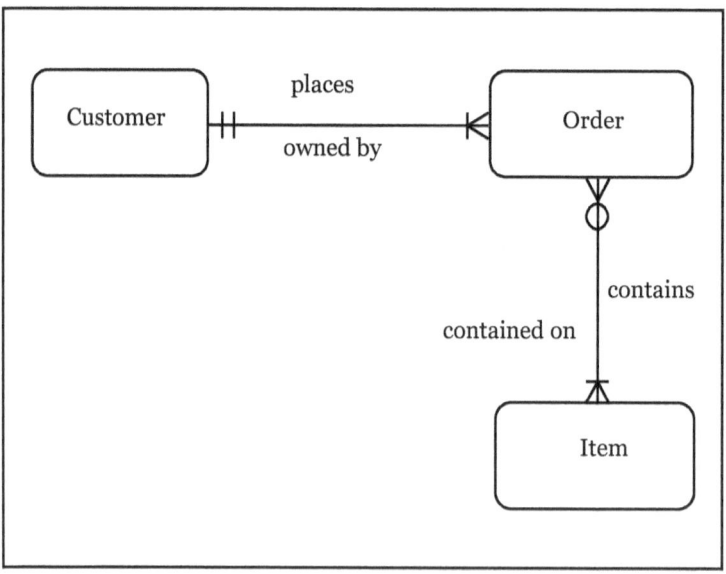

Events: events are identified or confirmed from the primitive entities and relationships.

- o customer places order
- o customer requests item availability
- o customer requests order status

Note: in this book, data model relationships are labeled in a clockwise fashion. See examples above.

Chapter 10
Business Event List

*The event list is a product of the
business team and defines relevant
business activities to which the
proposed system must respond. It is
the beginning of system requirements
and produces the system building
blocks. Each event is the beginning of a
common Thread that extends the
length of the lifecycle.*

The first serious attempt at defining functional requirements will be a list of business events to which the proposed system must respond. This list will define system scope at an upper level and will partition the proposed system into many relatively small, loosely coupled subsystems.

The event list is elicited from the business team and documents all business activities, one event at a time, that require a response from the proposed system. The event discovery task provides the opportunity for the business team to define, in terms of their natural business activities and in the business language, what the proposed system must do. This is a starting point for requirements definition. Later in the lifecycle, each of the system responses to these events will be specified in detail. The system response is the conceptual beginning of the requirements and will be defined in Use Cases.

The event is the Thread that will unify down-stream lifecycle components. From very early in the project to deployment, an event partitions and persists. An Event Thread originates in the event list but high level beginnings are recognizable in the Domain Model.

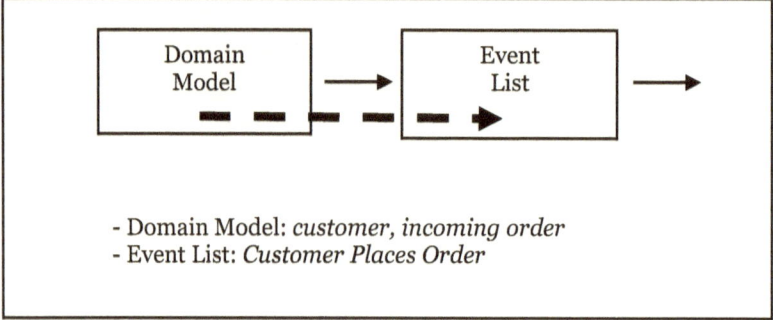

- Domain Model: *customer, incoming order*
- Event List: *Customer Places Order*

Procedure Overview

Meet with the business team in joint requirements sessions to identify the events from the business activity to which the proposed system must respond.

o   Use the context diagram to guide the early discussions. Walk through a typical day of business activity in the context of the system to be developed. *Storyboarding can be integrated into this process.*

o   The business team needs to look at their every-day activity and identify all work that will interface with the system. To get things going, the IT team can have a few examples in mind from their review of the system objectives and the current business activities.

o   Document these activities in the normalized form subject/verb/object, where the subject is the business role interfacing with the system and the object is the trigger (the input that will provoke a response from the system). Temporal events are typically documented as "Time to *verb/object*."

o   Identify the role (rather than an individual person). In the event *Customer Places Order*, the role is customer and the trigger is *order*.

From the case study

A partial list of events follows.

o   Customer Places Order

o   Customer Requests Item Availability

o   Customer Requests Order Status

o   Time to Generate Weekly Sales Report

o   Management Submits Item Discount Data

o   Customer Returns Item

Entity Lifecycle

"A state diagram relates events and states. When an event is received, the next state depends on the current state as well as the event; a change of state caused by an event is called a transition. A state diagram is a graph whose nodes are states and whose directed arcs are transitions labeled by event names. "The state diagram specifies the state sequence caused by an event sequence. If an object is in a state and an event labeling one of its transitions occurs, the object enters the state on the target end of the transition." [Rumbaugh, 1991]

An entity's lifecycle is the set of state transitions that an entity experiences from its inception through its ultimate obsolescence. The study of an entity lifecycle is the documentation of the possible relevant states and the events that move the entity from one state to another. This documentation is typically in the form of a diagram (see diagram in Procedure Overview below).

Early examination of an entity lifecycle can help discover events in the business area or can provide an understanding of the interaction between the entities and the events when used late in the analysis phase. An entity state represents a time window in the life of an entity. States within a single Entity Lifecycle Diagram must be mutually exclusive. Often, more than one lifecycle will exist for a single entity. For our order processing system case study, the ELCD reveals many events that have not been documented so far. It should be obvious to the development team that it has some work to do.

Procedure Overview
- ○ With help from the user, list the states of the system's major data entities (see example below).
- ○ For those data entities with a rich set of states, build an Entity Lifecycle Diagram (ELCD). An ELCD represents the sequence of valid entity states, depicts the dependencies between states, and documents the business event that causes a change in state.
- ○ Compare for completeness the events from the ELCD (example this section) with the event list developed early in the analysis phase.

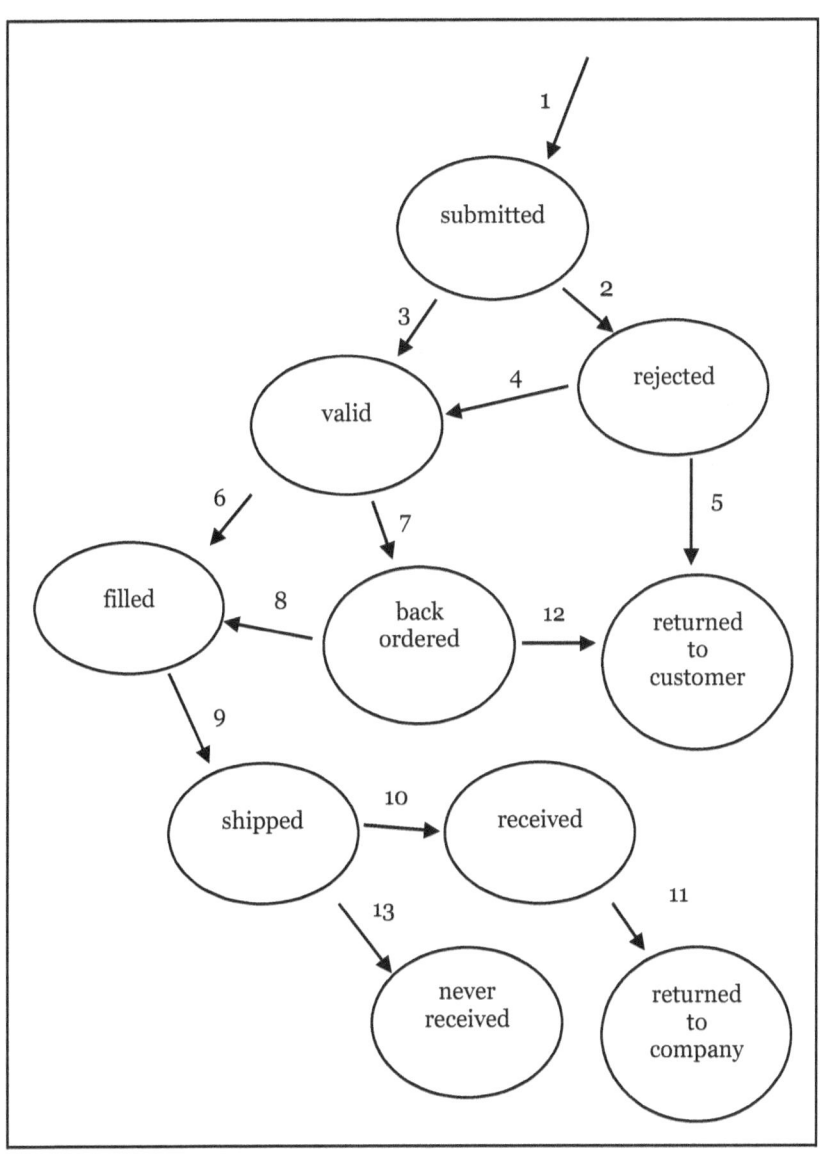

From the case study

Entity states of an order:

- submitted
- rejected
- valid

- filled
- back ordered
- shipped
- received
- returned to customer
- returned to company
- never received

Event candidates: *

1. customer submits order
2. order is rejected
3. order is validated
4. order is corrected
5. rejected order is returned to customer
6. order is filled
7. order is back ordered
8. back order is filled
9. order is shipped
10. order is received by customer
11. customer returns order
12. back order is returned to customer
13. shipped order is not received

* this is a discovery tool – the event candidates are not yet in the normalized event format (object/action/object).

System Response Table

The events occur in the business space, outside of the system. Requirements defined in the Use Cases are inside the system. At some point a transition from an outside perspective to an inside perspective must occur.

An abstraction of the processing that will take place when each event instance occurs is documented by the System Response Table (SRT).

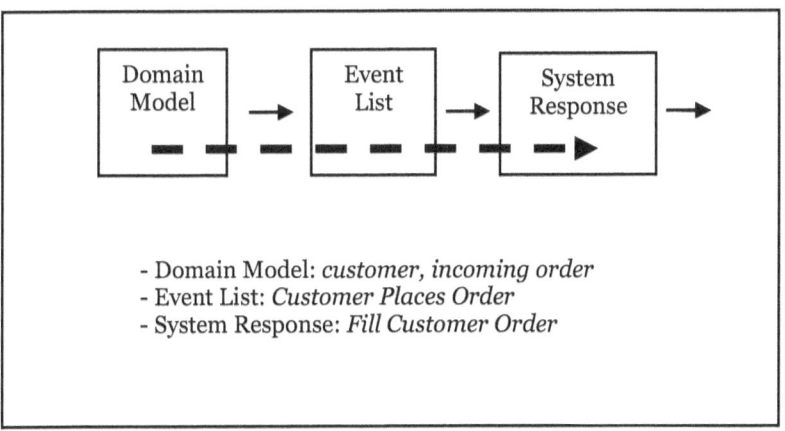

- Domain Model: *customer, incoming order*
- Event List: *Customer Places Order*
- System Response: *Fill Customer Order*

This table is a collection of system responses to the business events in the event list and renames the events to a system perspective (fourth column below). Data other than that shown below can be included in the table at the discretion of the development team (example: 'major outputs', 'output destination', and 'deployment priority').

Business Event	Source	Trigger	Event Response
Customer places order	Customer	Order	Fill customer order
Customer requests item availability	Customer	Item availability request	Return item availability
Customer requests order status	Customer	Order status request	Return order status
Time to generate weekly sales report	NA	(temporal)	Generate weekly sales report
Management submits item discount data	Management	Discount data	Apply item discount
Customer returns item	Customer	Returned item	Return item

Chapter 11
High-Level Event Use Case

*The requirements for each event are
described in a Use Case. A first-pass
high-level Use Case provides a starting
point for requirements definition and a
foundation for subsequent Event Use
Case detail. This event 'story' begins to
define requirements for each Thread.*

There is a lean Use Case for each event and associated system response. The event instance invokes the system response and the lean Use Case begins to define the functional requirements for the system response.

The lean Use Case documents our understanding of the Use Case goal at a high level. It is not adequate for the build phase but provides a catalyst for the early conceptual phase discussions and a basis for subsequent detail of the processing steps needed to produce the required output products.

Lean Use Case. A brief textual description made up of the Use Case name, user interface, major outputs, and description of the work to be completed (the Use Case goal).

The exercise of developing lean Use Cases provides for a deeper understanding of each system response and will give the teams a chance to agree on what in general each system response should be. It might also expose the need for event priority changes. As the end of the waterfall phase approaches, we have information on the number of Use Cases and some idea of their complexity and priority. *With this additional data, the development team should assess earlier decisions about schedule, resources, and scope.* This effort ends the waterfall phase of the lifecycle and moves to the iterative, just-in-time phase for a Use Case-by-Use Case development approach.

The event is the Thread that unifies the system response and the lean Use Case. It continues to tie the lifecycle layers together as the event *Customer Places Order* has its partner Use Case *Fill Customer Order*.

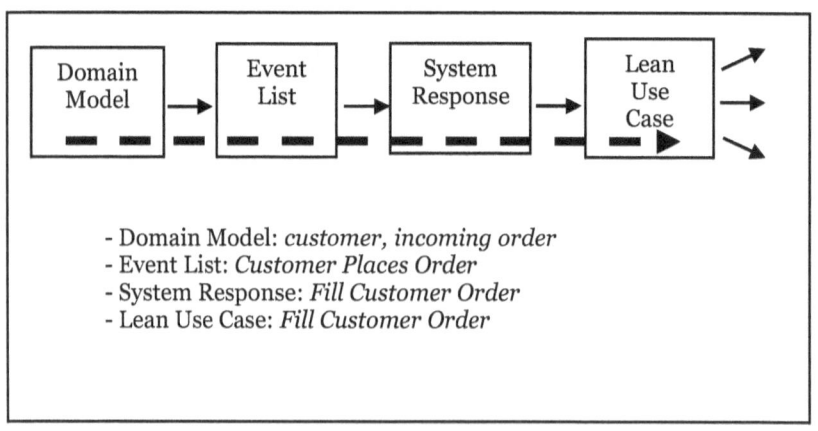

- Domain Model: *customer, incoming order*
- Event List: *Customer Places Order*
- System Response: *Fill Customer Order*
- Lean Use Case: *Fill Customer Order*

Procedure Overview

For each event response, develop a high-level Use Case. To define a lean Use Case, meet with the appropriate SME(s) on an event-by-event basis.

o Identify the actor(s) and the user interface.

o Document the work to be completed by the Use Case (Use Case goal).

o List the major outputs generated by the Use Case.

From the case study

Fill Customer Order is invoked by the customer when an order is submitted and the system must respond by filling that order. The order is a) verified and in-stock quantities are checked for sufficiency, b) payment data is verified and sent to credit card company, c) error conditions are reported with the option to correct when appropriate, d) pick list, invoice, and shipping documents are created, and e) data stores are updated.

Chapter 12
Event Use Case JIT Detail

*The requirements effort now shifts to
detail and the methodology from a
waterfall to an iterative, JIT strategy.
Detail Event Use Cases are developed
and sent down stream as they are
completed. Each Thread extends from
the requirements to the design, code,
and test phases.*

A JIT Use Case is an expansion of a lean Use Case. The waterfall
phase has produced lean Use Cases, one for each business event
and associated system response. But not nearly enough detail has
been defined from which to build a system. The iterative phase
(see Appendix A) will address one system partition (Use Case) at
a time and document specifications sufficient to build, test, and
deploy each Use Case. The latest of functional requirements is
captured rather than building to requirements that were
generated months before the build. With a JIT strategy, the
requirements are fresh for the build.

The steps for the agent/system interaction required for the Event
Use Case goal to be completed are described in a detailed Use
Case. Once invoked, a Use Case runs until its goal is met. For the
event *Customer Requests Order Status*, the Use Case describes
the user/system interface necessary to identify the customer and
order and to return the order information. It also specifies what

role the system must play in fulfilling the primary scenario and the exception paths of 'no order match' and 'customer not on file'.

The Use Case detail is completed just in time to be passed on to the build steps. And the Thread continues as the build, test, and deploy efforts focus on one event at a time.

The event instance invokes the system response and the JIT Use Case defines the functional requirements for the system response in sufficient detail from which to build. The event is the Thread that unifies the system response and the lean and detailed Use Cases. *Each event can be traced upstream to the earliest artifacts and will be traceable downstream all the way to deployment.*

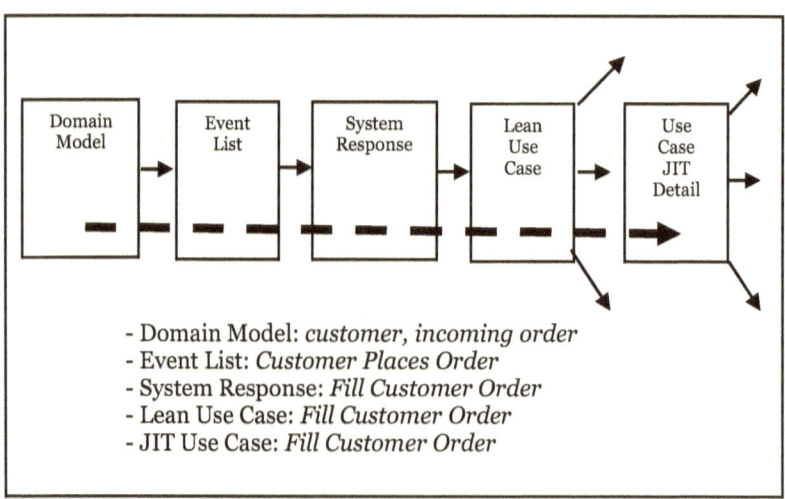

- Domain Model: *customer, incoming order*
- Event List: *Customer Places Order*
- System Response: *Fill Customer Order*
- Lean Use Case: *Fill Customer Order*
- JIT Use Case: *Fill Customer Order*

In parallel with Event Use Case development, the high-level data model is expanded to include all entities, relationships, and attributes on an event-by-event basis. Required data access defined in a Use Case can be used to test the data model to assure complete coverage before the database is developed.

Procedure Overview

The goal is to document requirements sufficient to build, test, and deploy each Use Case. The Use Case is made up of all the system and user steps necessary to fulfill the Use Case goal, that is, to produce all of the required output.

o Meet with the appropriate business SME(s) on a Use Case-by-Use Case basis.
o Refer to the Use Case diagram and the Lean Use Case description.
o Identify the event trigger that activates the Use Case.
o Define the user/system interface by listing the data elements that pass between the user(s) and the system.
o Document each step that the user and system must carry out to complete the primary path towards the Use Case goal.
o Describe all alternate successful paths through the Use Case.
o Describe all exception (unsuccessful) paths that can occur in the operation of the Use Case.
o As the above steps are completed, identify and document all input and output data elements.

From the case study

Refer to Chapter 16 for the detail Use Case for *Fill Customer Order*.

Chapter 13
The Event Advantage

An event-driven approach to software development partitions the proposed system as natural business building blocks are identified early. The Event Thread stretches from requirements to deployment providing an adaptive, resilient environment.

Adopting an event-driven strategy brings with it some powerful advantages. The system is partitioned into highly independent pieces and each partition is a natural business activity that will continue to be recognized throughout the project. The early conceptual tasks are business driven and the business events drive the key system artifacts.

But even more powerful are the Event Threads that have their origin in the events and extend through the lifecycle from beginning to end. As work passes from one lifecycle step to another, each Event Thread continues to be a common thread tying the lifecycle steps and artifacts together.

Partitioning

The objective of system partitioning is to divide both the conceptual and physical system into 'chunks.' The partitioning scheme should produce independent subsystems that will support subsequent lifecycle efforts.

Partitioning of the proposed system space is a major outcome of the early conceptual phase.

Partitioning allows the development teams to work on independent pieces of the proposed system. The partitions simplify project, change, and risk management tasks since each partition is a virtual subsystem to be managed separately.

Events

Events are natural business activities familiar to the business team. And they become the natural building blocks of the proposed system. When a system is partitioned using business events, the beginning is somewhere in the middle of the hierarchy with familiar user activities (events). The model follows a process referred to as "middle out"; it is synthesized upward and decomposed downward as needed.

The event partitions persist from the requirements phase through construction, testing, and deployment. They can be managed and developed independently based on data dependencies and priorities. Change, risk, and other project management elements benefit from being able to focus on relatively small independent pieces of the system.

The Thread

Events divide the proposed system space into subsystems. But what occurs subsequently is even more powerful. Each event and system response pair retains its identity throughout the lifecycle;

this becomes the **common Thread** that brings the lifecycle phases and tasks to a single focus and ties the development artifacts together for each partition. A particular event can be found in each lifecycle phase and artifact extending into the production environment.

An event and its associated system response identified in the initial discovery JAD sessions establish a Thread that passes through requirements, design, build, test, UAT, and deployment. An Event Thread slices through the entire development process.

Event Use Cases

An event-driven Use Case is one that is derived from a *business event* (previously defined in the Event List). Each event has a system response and each system response is defined by a Use Case. The advantage of defining business events first is that they occur in the business space and are close to the users' natural everyday activity.

Since requirements begin to age soon after they are defined, to build from requirements that are months old increases the odds of producing a system that does not meet business needs. Defining and verifying Use Case requirements just-in-time allows delivery of a current system. In an event strategy, change has manageable impact because relatively small, isolated Event Threads are dealt with.

Events and Methodology

The domain model is the artifact that kicks off the system development effort. It establishes the roots of many of the events. The event list is the artifact that begins to look at the system in detail by defining the required business activities to which the system must respond. The event list partitions the system and establishes the common threads that will link the early and late lifecycle efforts. The acceptance plan and the deployment plan are also event driven.

Once the event building blocks are identified, Use Cases define the functional requirements for each Event Thread and provide the just-in-time specifications for the subsequent build, test, and deploy efforts.

The Advantage

So why use business events as a partitioning scheme?

1. *Events are adaptive.* Since an event response is tightly focused on a single response to a business activity, it is easier to isolate changes to a partition of specifications or a partition of code resulting from a change in the system requirements. This facilitates assessment of the impact and the subsequent implementation of the change. It is also easier to identify and mitigate risk when focusing on a single event partition.

2. Events promote involvement by the user (business team) since they are described in the *users' language* and since they

describe the users' work activities. They involve the user early and throughout the development lifecycle.

3. Events are natural business partitions which are established in the user interface and are the origin of common lifecycle Threads. These partitions continue to be *recognizable* throughout the development process, aiding in traceability from code back to requirements and facilitating communication between the development and user teams.

4. Event responses are *highly independent* subsystems (have very low coupling) that interface only with the users and stored data. By definition, they do not interact directly with other event responses. Also, for many business applications, a high percentage of these partitions are relatively simple custodial accesses or updates to the database.

5. Early *verification* of requirements and user acceptance testing can be accomplished on an event-by-event basis thus partitioning these tasks into manageable pieces.

Events are the outcome of the early conceptual phase and become the building blocks of the system. Each event is the beginning of a common thread that extends through each lifecycle step and artifact and establishes user familiarity and traceability throughout the project. And events are adaptive, allowing the project teams to respond to the natural changes in the business that will surely occur during the life of the system development effort.

Chapter 14
Physical Phases - Brief Discussion

*An event-driven approach to software
brings with it many advantages as the
software effort moves to the physical.
These tasks can be approached on an
event-by-event basis as building blocks
are prepared for deployment.*

The conceptual models represent 'what' the proposed system should do. The conceptual phases produce a system model that partitions the system with natural events from the business space. Each event retains its identity across the lifecycle and provides a common Thread for the entire development effort. Along with a data model, the conceptual tasks produce requirements defined in Event Use Cases which are passed to design as they are completed and verified just in time (JIT).

The physical phases establish 'how' the system will be built and translate the abstract conceptual model into the specific technical design for the new system. The common Event Thread affects the physical tasks as well as the conceptual model. Tasks such as software design, code, and test are completed on an event-by-event basis. User acceptance tests can be conducted on a single event and the event building blocks deployed one by one.

The physical phases accomplish:

- o   database design
- o   software design
- o   GUI
- o   build
- o   component and system test
- o   UAT
- o   database population
- o   component release

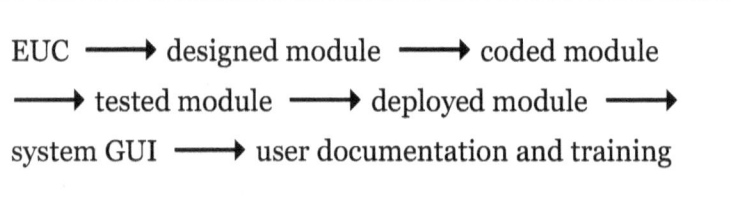

The Thread extends from the conceptual into and through the physical phases, tracing the physical tasks back to their roots in the conceptual model. Even the user documentation and training benefit from the event partitioning (see table below).

	Customer Places Order	Customer Requests Item Availability	Customer Requests Order Status	Time to Generate Weekly Sales Report	Management Submits Item Discount Data	Customer Returns Item
Database Structure						
Design						
Build						
Component Test						
System Test						
UAT						
Database Population						
Component Release						
User Training						

These physical tasks can be approached on a component-by-component basis, in sync with and dependent upon the database being built and following the Event Use Case priority. System subsets of these component building blocks can often be implemented as a prototype or as a production app to be used to populate the database.

When the physical phases follow a strong, event partitioned, conceptual effort, the system being built and the software team become more agile and more resilient.

Coding

I remember from my early days in the 70s as a programmer for a contractor at the space center in Houston. I had been given a Fortran program that someone else had written and subsequently left the company. It was filled with the notorious 'GO TO' statement. I was struggling to make changes to the program on a manned mission schedule. I finally stretched the printout down the hallway, crawled on my hands and knees and drew a line from each 'GO TO' to its target code, trying to get some sense of the program structure. This was a classic example of 'spaghetti code' and it took its toll. How nice it would have been to work on one event at a time.

Event partitioning and the Thread changes all of that. By the very nature of events, each coded program is a partition within the whole system and is typically a relatively small one, with low coupling and high cohesion. This partitioning facilitates the more structured approach to coding typically seen today. This code has traceability back to requirements and back to the very business activity that initially identified it as a part of the system. It traces forward to tests and deployment efforts, never losing the identity that was established early in the requirements phase.

Events establish system architecture but also allow coding to focus on a concise piece of function.

Event-Driven Testing

Unit, integration, and system level testing do not verify the correctness of system requirements. These tests just verify that the requirements, correct or incorrect, were developed as specified and that the software runs without errors. The only test designed to verify requirements is the user acceptance test (UAT). UATs are based on the requirements and check to see that the correct solution from a business perspective was developed.

It's widely accepted in the IT industry that testing should begin as early as possible. Generating test cases from Event Use Cases facilitates early testing and also partitions the testing effort for an effective methodology. Test cases can be created as soon as a Use Case is completed and manually verified, before any code is written. This partitions the test effort on an event-by-event basis and involves the test team earlier in the development effort. The Thread cuts through the testing effort as it continues on through the lifecycle.

Test Case

A test case is a set of input data, execution procedures, and expected results developed for a single Event Use Case scenario (a specific requirement). The purpose of a test case is to partition the testing effort and to specify test conditions necessary to verify a particular path through a Use Case. Each Scenario needs a test case and more likely will need more than one as valid and invalid data are tested along with data just inside and outside boundaries.

The relationship of test cases to Use Cases is represented in the decomposition below.

Use Case 1
    Scenario 1: Test Case 1
        a.   input data (actual values)
        b.   execution procedures
        c.   expected results

Scenario 1: Test Case 2

    a.   input data (actual values)

    b.   execution procedures

    c.   expected results

Scenario 2: Test Case 1

    a.   input data (actual values)

    b.   execution procedures

    c.   expected results

Scenario n: Test Case n

    a.   input data (actual values)

    b.   execution procedures

    c.   expected results

Input Data

Actual data values are provided in the script so that the expected results can be predicted character by character. It is necessary to provide values just inside and just outside of the valid data boundaries as well as valid values that fall well within the range and values that fall well outside the valid data set. Data values that represent special cases gleaned from the business rules should also be tested.

Execution Procedures

Procedures outline a step by step execution of the test case including the entry of data values and the selection from presented GUI options. These steps leave no room for variation and precisely lead the tester through the test case.

Expected Results

Based on the data provided in the procedures, the expected results can be predicted and any variation is noted in the test case notes (along with nominal results).

Database Test Data

Preparing a test database is typically a sizable task. Data must be available for all scenario test cases, and there can be many values per scenario. For example, products need quantity on hand that forces both sufficient and insufficient conditions. And it is common to have more than one value for each condition. When testing for a quantity ordered of "3", sufficient database quantities such as "4", "100", and "1000" could be used; zero and "2" would test the insufficient conditions. A test database is built in parallel with the test cases as the needed data is identified and defined.

Early Testing

Use Cases are associated with early conceptual phases. Using them to drive testing involves the test team much earlier and identifies problems much earlier in the lifecycle. Using Use Cases and associated test cases as a basis for testing simplifies the testing effort. It also increases efficiency and helps to insure complete test coverage.

Book 2 Epilogue

Book 2 has explored the application of the Event Thread. A two-phased methodology takes a high-level look at the proposed system then drills deeper to provide detail sufficient for the build phase.

This mix of waterfall and iterative strategies brings the best of each to the software development effort. Requirements begin to age virtually immediately after they are documented. Tackling a complex problem at a high level in the beginning gives the business and IT teams the best chance of establishing a strong foundation for the software development task. A subsequent iterative strategy produces fresh requirements for downstream design and development by partitioning the work Use Case by Use Case.

All the while the Event Thread gives the teams common threads to follow throughout the complex development lifecycle and into the production phase.

Book 3
Case Study

The running case study is a simple merchandize order system such as might be found on the Web. In addition to order-related transactions, it allows for the return of an item, management control of discount information, and generation of various reports.

Book 3 includes the artifacts for all six case study events as well as three test scenarios for the event *Customer Places Order.*

This case study is included to demonstrate the physical application of the concepts and methods of Books 1 and 2. *It is not intended to be a complete solution.*

Chapter 15
Case Study Artifacts

*Various artifacts are created as the
system development lifecycle proceeds
from requirements definition to
deployment. Information about the
proposed system ranges from high
level early in the process to detail as
the effort drills down deeper and
deeper. The artifacts that emerge and
are to be documented are defined in
this chapter.*

Domain Model

The resulting model provides an overall framework in which to
add business events. A context diagram and high-level data model
make up the domain model (see chapter 9). Domain modeling
provides a starting point for discussion between the business and
IT teams preceding requirements definition. It also helps identify
who should be present at subsequent joint requirements
gathering sessions. And the business events have roots in this
model.

Context Diagram

Building the Context Diagram consists of exploring and
explaining the environment of the problem to be solved and its
interfaces to the proposed system. Event partitioning begins as
data flows spawn events in the subsequent event list. Each Event
Thread has its beginning in this model.

Lean Data Model

The Domain Model includes a Lean Data Model and documents data entities and data relationships at a high level. As entities and relationships are identified, more potential business events are revealed. Events will in turn lead to new entities and relationships. At this time, the business rules might not be known adequately to set cardinality and resolve many-to-many relationships. As we progress through event discovery, the data model will become more complete as it is normalized. Refer to the normalized data model in Appendix D.

Event List

The event list documents all business activities, one event at a time, that require a response from the proposed system. Events occur outside the system. The event discovery task provides the opportunity for the business team to define, in terms of their natural business activities and in the business language, what the proposed system must do. This is a starting point for requirements definition.

As new events are defined, normalized components of the data model emerge as do new components of the Context Diagram. Similarly, discoveries in the domain model often reveal new events.

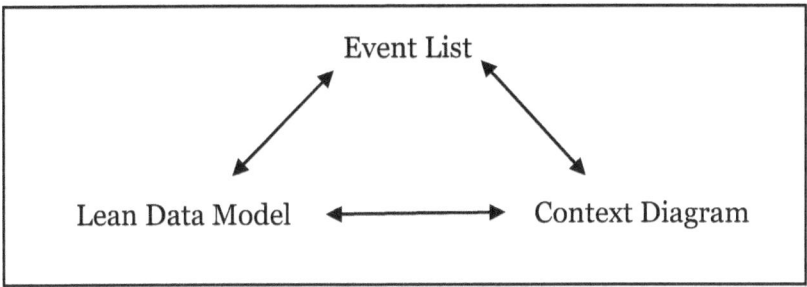

System Response

Each business event has a corresponding system response. This system response conceptually represents, as a black box, the system requirements to be subsequently defined. It is simply a renaming of the business events to a system perspective (for example: *Customer Places Order* renamed to *Fill Customer Order*).

Lean Use Case

There is a lean Use Case for each event and associated system response. The event instance invokes the system response and the lean Use Case begins to define the functional requirements for the system response.

The lean Use Case documents our understanding of the Use Case goal at a high level. It is not adequate for the build phase but provides a catalyst for the early conceptual phase discussions and a basis for subsequent detailing of the processing steps needed to produce the required output products.

It provides a brief textual description made up of the Use Case name, user interface, major outputs, and textual description of work to be completed (the Use Case goal).

Use Case Diagram

The Use Case diagram is developed in parallel with the lean Use Cases. It provides a visual representation of the Use Cases and the system interface to the actors. The role of each actor and the complexity of the actor role can be seen.

Use Case JIT Detail

1. Use Case name
2. Use Case ID, author, creation date, system name
3. Priority
4. Use Case goal: (one sentence)
5. Use Case summary: processing plus input and output
6. Use Case Trigger: business event name and data flow
7. Actors
8. Required preconditions
9. Primary Scenario
10. Alternative Paths
11. Expected Post Conditions
12. Business Rules
13. Input Summary
14. Output Summary
15. Use Case Issues
16. Use Case Notes
17. Non-Functional Requirements

Requirements Verification

When possible, errors and omissions need to be discovered during the conceptual modeling phase when the cost of repair is the lowest. With an event strategy and the associated Threads that occur, verification of the requirements can focus on a single Use Case and on the associated user acceptance test that will ultimately be used for final confirmation of the correct build. A SME and the appropriate requirements analysts can isolate an effort to carefully examine the contents of a single event Use Case and to confirm or repair the requirements. Changes at this point are many times less costly to make than those detected later in the lifecycle.

Design, Build, Test

Within an architectural structure for the system under development, each Use Case can be designed and built and unit tested as a separate subsystem ready for integration and system testing. Focus for this effort can be on a single module reducing the overall complexity of the design and unit testing tasks.

User Acceptance Test

Unit, integration, and system level testing do not verify the correctness of system requirements. These tests just verify that the requirements, correct or incorrect, were developed as specified and that the software runs without errors. The only tests designed to verify requirements are the user acceptance tests (UAT). These tests are based on the requirements and check to see that the correct solution from a business perspective was developed. The problem with discovering errors and omissions at this point in the development lifecycle is that UAT occurs after development and is much too late. But these tests are nevertheless necessary to identify problems not found in the verification phase. Once the Use Case is verified and built, the Use Case software can be user tested and either repaired or prepared for deployment.

System Test, Deployment

Following the priority established in the Detail Event Use Case, each Use Case can be system tested and readied for deployment. Although typically systems aren't deployed Use Case by Use Case, it's possible that one or more subset of the total set of Use Cases can return value from an early deployment.

Chapter 16
Fill Customer Order

*Fill Customer Order is initiated by the
customer when an order is submitted.
The system must respond by filling the
order.*

Context Diagram

The Event Thread for **Fill Customer Order** has its beginning in
this model as the customer submits an order for an item.

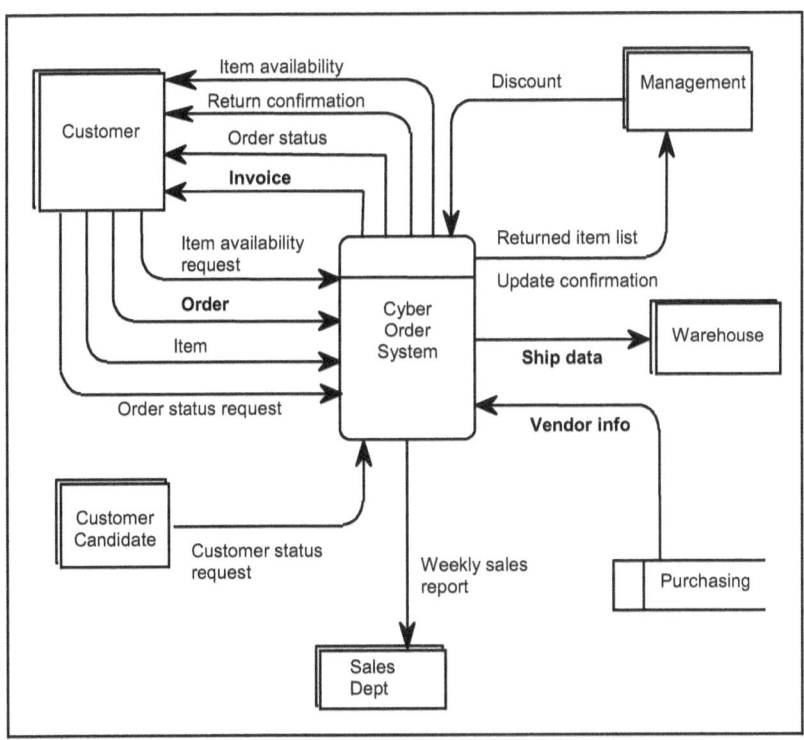

Lean Data Model

As entities and relationships are identified, more potential business events are revealed. The data model fragment must support an association between customer and item. The order data entity emerges as a business entity in its own right, but also will act to resolve the many-to-many relationship between the customer and available items. The many-to-many relationship between Order and Item must eventually be resolved to hold item data for an order.

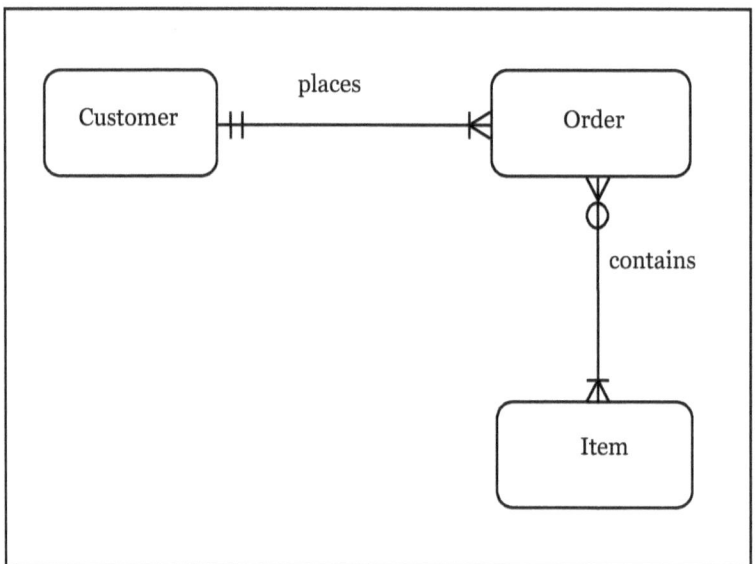

Event List

1. ***Customer Places Order***
2. Customer Requests Item Availability
3. Customer Requests Order Status
4. Time to Generate Weekly Sales Report

5. Management Submits Item Discount

6. Customer Returns Item

System Response

1. ***Fill Customer Order***

2. Return Item Availability

3. Return Order Status

4. Generate Weekly Sales Report

5. Apply Item Discount

6. Return Item

Lean Use Case

Fill Customer Order is initiated by the customer when an order is submitted and the system must respond by filling that order and creating shipping artifacts.

Use Case Diagram

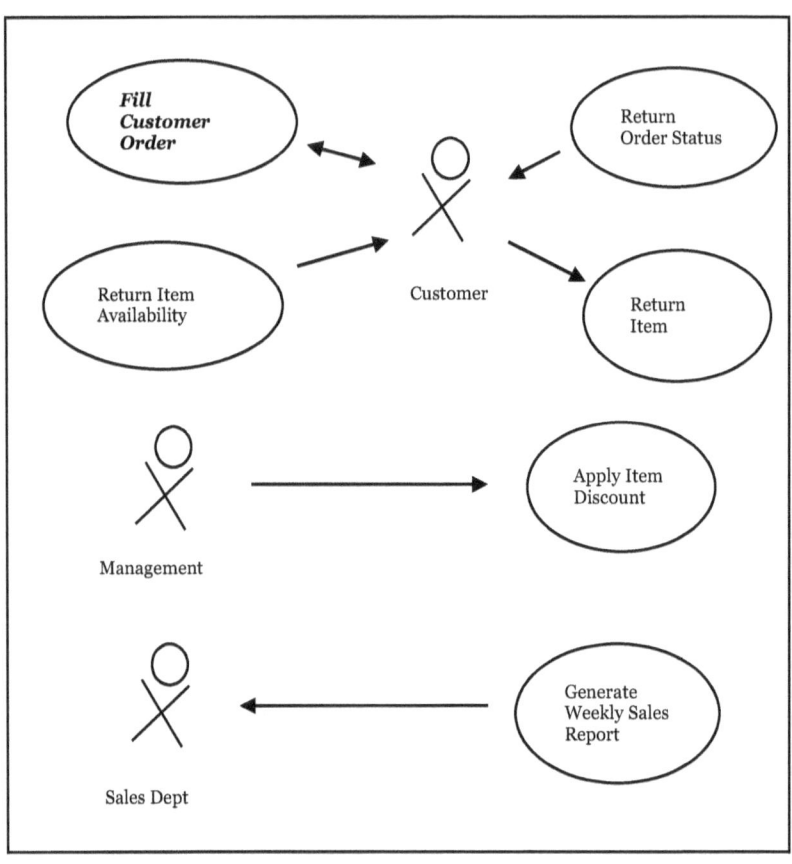

Use Case JIT Detail

1. ***Fill Customer Order***

2. UC # 1; Wiley; created November 2009; Web Order System (WOS)

3. Priority: 1

4. Goal: to accept an order from a customer and fill it with all the items requested.

5. *Fill Customer Order* is initiated by the customer when an order is submitted and the system must respond by filling that order. The order is:

1) verified and in-stock quantities are checked for sufficiency,
2) payment data is verified and sent to credit card company,
3) error conditions are reported with the option to correct when appropriate,
4) pick list, invoice, and shipping documents are created,
5) data stores are updated.

6. Use Case Trigger (business event): *Customer places order* (order)

7. Actors: customer, credit card company, warehouse

8. Preconditions:

1) customer must be on file

9. Primary Scenario: Successful Order

1) select 'place order' from menu
2) customer submits order information
3) system validates order information
4) system checks for sufficient quantity of product
5) system requests pay information from customer
6) customer submits pay information
7) system exports pay data
8) system updates data stores
   a) quantity on hand
   b) order/item instances created
   c) order status set to "filled"
9) system generates pick list, invoice, shipping documents
10) system informs customer of successful order completion

11) customer closes order session

10. Alternative Paths

Alternative Path 1: Insufficient Item Quantity

1) leave primary path following step 4
2) system reports insufficient item quantity
3) system offers backorder option
4) customer accepts partial order / backorder option
5) system creates backorder
6) return to primary path step 5

Alternative Path 2: Out-of-stock Condition

1) leave primary path following step 4
2) system reports out-of-stock condition
3) system offers backorder option
4) customer submits backorder choice
5) system creates backorder
6) system updates data stores
7) system presents backorder data to customer
8) return to primary path step 5

Alternative Path 3: Incomplete Order Data

1) leaves primary path following step 3
2) system reports invalid order data
3) system offers corrective action
4) customer submits corrected information
5) return to primary path step 3

Alternative Path 4: Invalid Credit Card

1) leave primary path following step 7
2) system reports invalid credit card
3) customer terminates session

      4) system generates credit card warning as prescribed by law

11. Post Conditions:
    1) data stores updated with shipping, inventory, and payment data
    2) customer has shipping and tracking data
    3) credit card company has order charges

12. Business Rules:
    1) invalid credit card will be reported within 1 hour
    2) order will not be filled if order amount takes card over limit
    3) backorder will be shipped within 5 business days
    4) orders out of stock will be offered substitute products
    5) orders will not be delivered to post office box
    6) credit card orders only will be accepted
    7) orders will ship only to credit card address

13. Input Summary:
    1) name
    2) card address
    3) delivery address
    4) card number
    5) card security number
    6) order item number
    7) item quantity
    8) vendor data

14. Output Summary:
    1) order confirmation
    2) pick list
    3) shipping manifest

4) shipping label

5) updated data stores

15. Use Case Issues/Risk:

    1) backorder policies

    2) how/who to notify of invalid credit card

    3) credit card interface

16. Use Case Notes:

    1) law for invalid credit card needs to be researched

    2) new customer will be processed in an Extend Use Case

    3) product substitution algorithm not developed

    4) order status is updated by another system when order is shipped

17. Non-Functional Requirements:

    1) security: credit card data will be passed along industry standard secure line

    2) performance: response to user will be less than 5 seconds

    3) design: Event Threads will remain independent and identifiable throughout the lifecycle

    4) reliability: system must be available to the customer 98% of the time (24/7)

Requirements Verification

Changes at this point are many times less costly to make than those detected later in the lifecycle. Once the Use Case definition is complete, the requirements for *Fill Customer Order* can be verified and the UAT can be designed.

Design, Build, Test

Focus for the design, build, and test effort can be on the single module **_Fill Customer Order_** reducing the overall complexity of these tasks.

User Acceptance Test

Once the Use Case is verified and the UAT written, the **_Fill Customer Order_** software can be user tested and either repaired or prepared for deployment.

System Test, Deployment

Following the priority established in the Detail Event Use Case, **_Fill Customer Order_** can be system tested and readied for deployment.

Chapter 17
Return Item Availability

Return Item Availability is initiated by the customer when a request for the availability of an item is made. The system must respond by returning the quantity available.

Context Diagram

The Event Thread for **Return Item Availability** has its beginning in this model as the customer submits a request for the availability of an item.

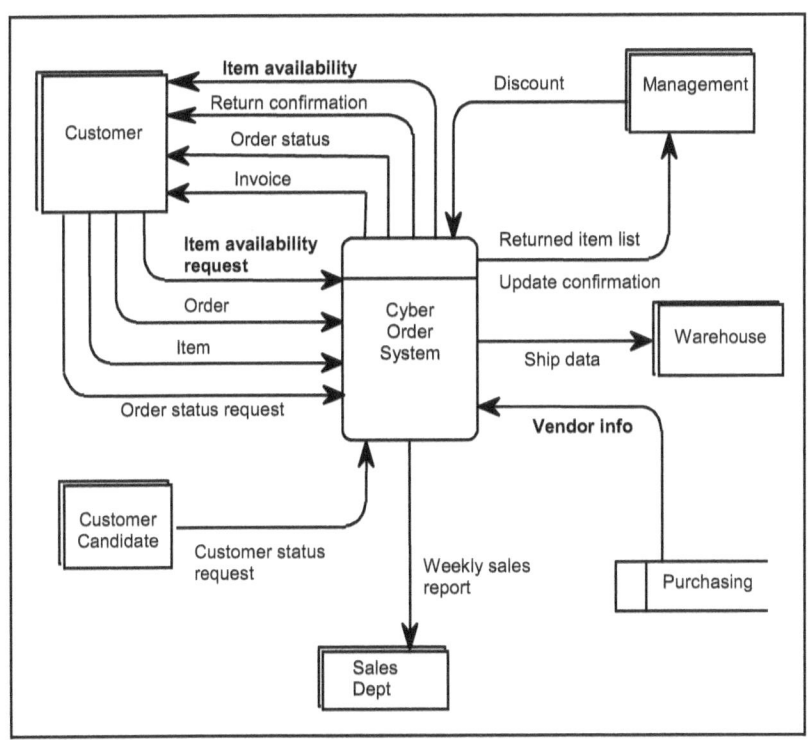

Lean Data Model

As entities and relationships are identified, more potential business events are revealed. The data model fragment must support identification of customer and retrieval of item quantity for the specified item.

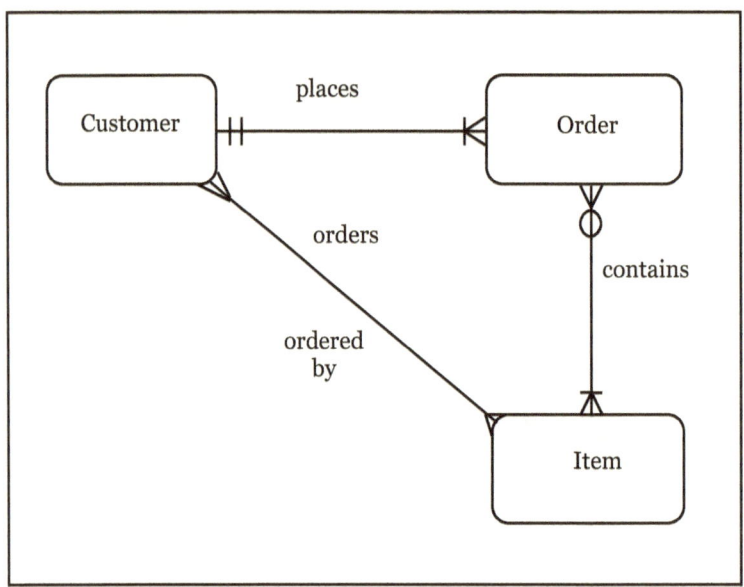

Event List

1) Customer Places Order
2) *Customer Requests Item Availability*
3) Customer Requests Order Status
4) Time to Generate Weekly Sales Report
5) Management Submits Item Discount
6) Customer Returns Item

System Response

1) Fill Customer Order

2) *Return Item Availability*

3) Return Order Status

4) Generate Weekly Sales Report

5) Apply Item Discount

6) Return Item

Lean Use Case

Return Item Availability is initiated by the customer when a request for the availability of an item is made. The system must respond by verifying the customer and item number and returning the quantity-on-hand and the quantity-on-order.

Use Case Diagram

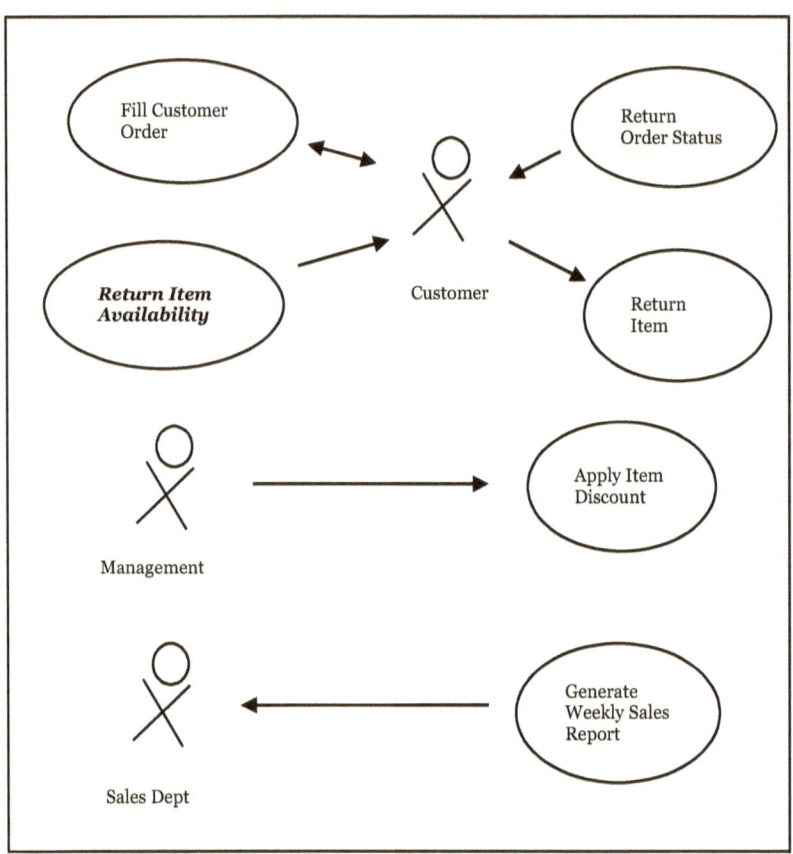

Use Case JIT Detail

1. ***Return Item Availability***

2. UC # 2; Wiley; created November 2009; Web Order System (WOS)

3. Priority: 4

4. Goal: to provide the quantity available and quantity on order for the specified item.

5. *Return Item Availability* is initiated by the customer when a request for the availability of an item is submitted. The system must respond by:

    1) verifying the customer and item numbers,

    2) rejecting an invalid item number with an option to correct,

    3) returning to the customer the quantity-on-hand and the quantity-on-order of the specified item.

6. Use Case Trigger (business event): *Customer Requests Item Availability* (item availability request)

7. Actors: customer

8. Preconditions:

    1) customer must be on file

9. Primary Scenario:

    1) customer selects 'request item availability'

    2) customer submits request

    3) system checks customer status

    4) system verifies item number

    5) system returns quantity-on-hand and quantity-on-order

    6) customer ends query session

10. Alternative Paths

    Alternative path #1: Customer not on file

        1) leave primary path following step 3

        2) system reports customer not on file

        3) system offers opportunity to establish customer profile

        4) customer submits customer profile data

        5) system validates customer profile

        6) return to primary path step 4

    Alternative path #2: Invalid item

     1) leave primary path following step 4

     2) system reports invalid item

     3) system offers opportunity to change item number

     4) customer submits another item number

     5) return to primary path step 4

11. Post Conditions:

     1) customer has item availability

     2) status of data stores unchanged

12. Business Rules:

     1) only customers on file can access item availability

     2) after two failed tries, session will be dropped

13. Input Summary:

     1) customer id

     2) item number

     3) vendor data

14. Output Summary:

     1) item number

     2) item description

     3) quantity-on-hand

     4) quantity-on-order

15. Use Case Issues/Risk:

     1) how many invalid item numbers to allow before dropping session

     2) how many invalid customer ids to allow before dropping session

16. Use Case Notes:

     1) none

17. Non-Functional Requirements:
    1) security: customer must be on file
    2) performance: response to user will be less than 5 seconds

Requirements Verification

Changes at this point are many times less costly to make than those detected later in the lifecycle. Once the Use Case definition is complete, the requirements for **Return Item Availability** can be verified and the UAT can be designed.

Design, Build, Test

Focus for the design, build, and test effort can be on the single module **Return Item Availability** reducing the overall complexity of these tasks.

User Acceptance Test

Once the Use Case is verified and the UAT written, the **Return Item Availability** software can be user tested and either repaired or prepared for deployment.

System Test, Deployment

Following the priority established in the Detail Event Use Case, **Return Item Availability** can be system tested and readied for deployment.

Chapter 18
Return Order Status

Return Order Status is initiated by the customer when a request for the status of an order is made. The system must respond by verifying the customer and order number and returning the order status.

Context Diagram

The Event Thread for **Return Order Status** has its beginning in this model as the customer submits a request for the status of an order.

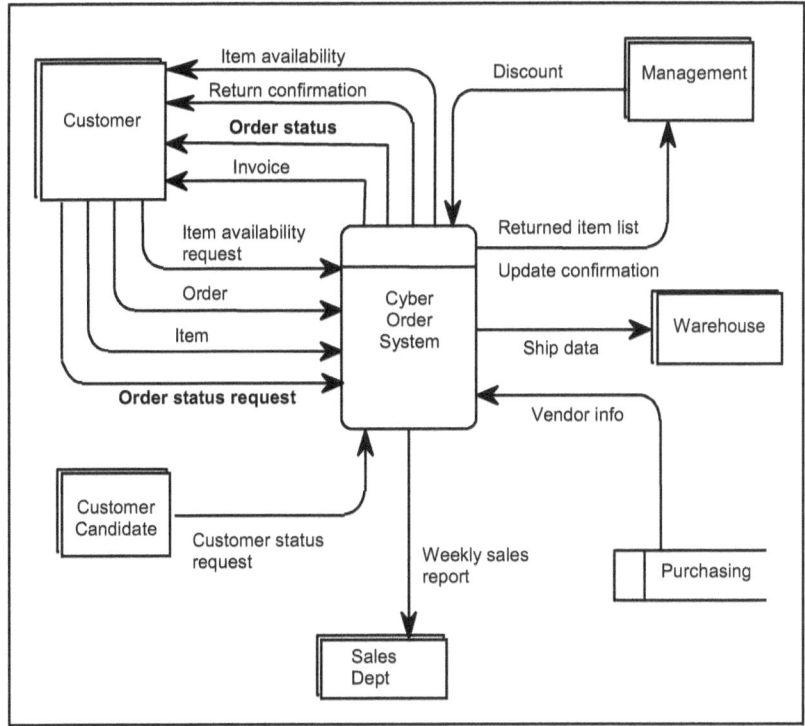

Lean Data Model

As entities and relationships are identified, more potential business events are revealed. The data model fragment must support verification of customer and order and the derivation of the status of items on order.

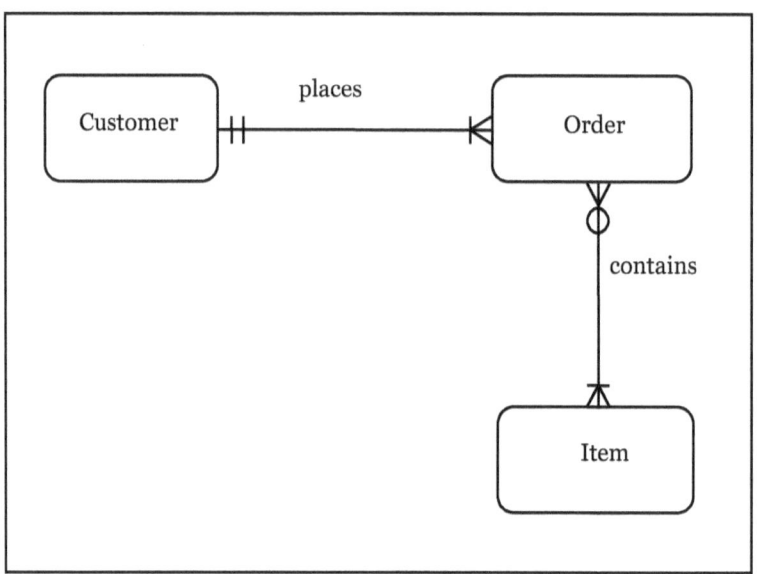

Event List

1. Customer Places Order
2. Customer Requests Item Availability
3. ***Customer Requests Order Status***
4. Time to Generate Weekly Sales Report
5. Management Submits Item Discount
6. Customer Returns Item

System Response

1. Fill Customer Order
2. Return Item Availability
3. ***Return Order Status***
4. Generate Weekly Sales Report
5. Apply Item Discount
6. Return Item

Lean Use Case

Return Order Status is initiated by the customer when a request for the status of an order is made. The system must respond by verifying the customer and order number and returning the order status.

Use Case Diagram

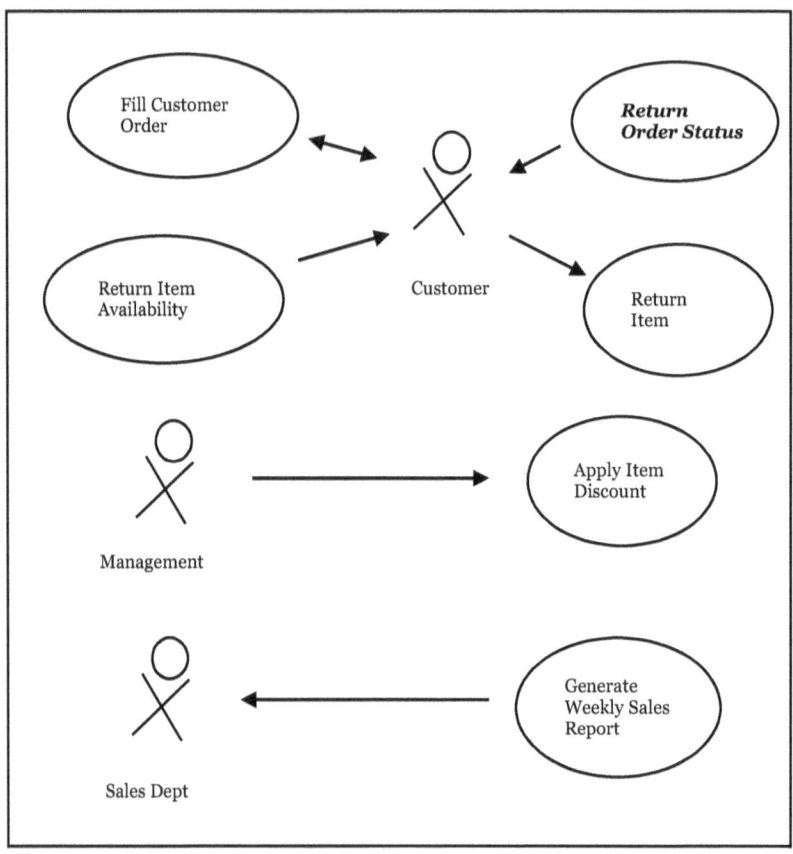

Use Case JIT Detail

1. ***Return Order Status***

2. UC # 2; Wiley; created November 2009; Web Order System (WOS)

3. Priority: 3

4. Goal: to provide the status of an order for a specified item.

5. *Return Order Status* is initiated by the customer when a request for the status of an order is submitted. The system must respond by:

1) verifying the customer and the order number,

2) rejecting an invalid customer or order number with an option to correct,

3) returning to the customer the status of the specified order.

6. Use Case Trigger (business event): *Customer Requests Order Status* (order status request)

7. Actors: customer

8. Preconditions:

1) customer must be on file

2) order must be on file

9. Primary Scenario:

1) customer selects 'request order status'

2) customer submits request

3) system checks customer status

4) system verifies order number

5) system derives order status

6) system returns status of specified order

7) customer ends query session

10. Alternative Paths

Alternative path #1: Customer not on file

1) leave primary path following step 3

2) system reports customer not on file

3) system offers opportunity to establish customer profile

4) customer submits customer profile data

5) system validates customer profile

6) return to primary path step 3

Alternative path #2: Invalid order

1) leave primary path following step 4

2) system reports invalid order

3) system offers opportunity to correct order number

4) customer submits corrected order number

5) return to primary path step 4

11. Post Conditions:

1) customer has received order status

2) data stores are unchanged

12. Business Rules:

1) only customers on file can access order status

2) after two failed tries, session will be dropped

13. Input Summary:

1) customer id

2) order number

14. Output Summary:

1) order number

2) order submit date

3) order status

4) order status by item

15. Use Case Issues:

1) how many invalid order numbers to allow before dropping session

2) how many invalid customer ids to allow before dropping session

16. Use Case Notes:

1) if order status in Order entity is not "filled", order status must be returned by item

17. Non-Functional Requirements:

1) security: customer must be on file

2) performance: response to user will be less than 5 seconds

Requirements Verification

Changes at this point are many times less costly to make than those detected later in the lifecycle. Once the Use Case definition is complete, the requirements for **Return Order Status** can be verified and the UAT can be designed.

Design, Build, Test

Focus for the design, build, and test effort can be on the single module **Return Order Status** reducing the overall complexity of these tasks.

User Acceptance Test

Once the Use Case is verified and the UAT written, the **Return Order Status** software can be user tested and either repaired or prepared for deployment.

System Test, Deployment

Following the priority established in the Detail Event Use Case, **Return Order Status** can be system tested and readied for deployment.

Chapter 19
Generate Weekly Sales Report

Generate Weekly Sales Report is initiated when a specified time is reached. The system must respond by reporting all sales for the previous week.

Context Diagram

The Event Thread for **Generate Weekly Sales Report** has its beginning in this model as the *time arrives* to generate the weekly report (temporal event).

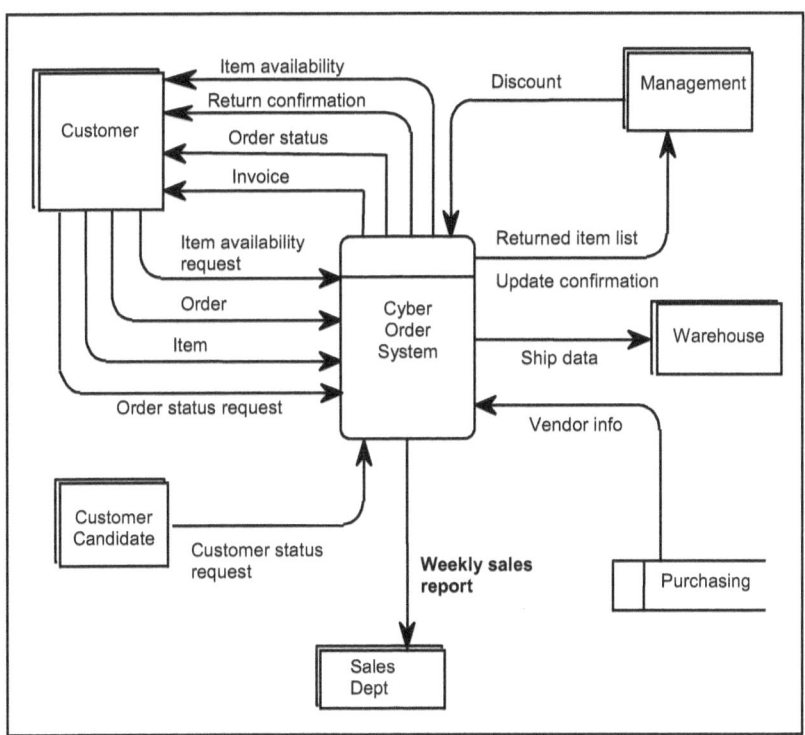

Lean Data Model

As entities and relationships are identified, more potential business events are revealed. The data model fragment must support retrieval of order and item data for a specified date range.

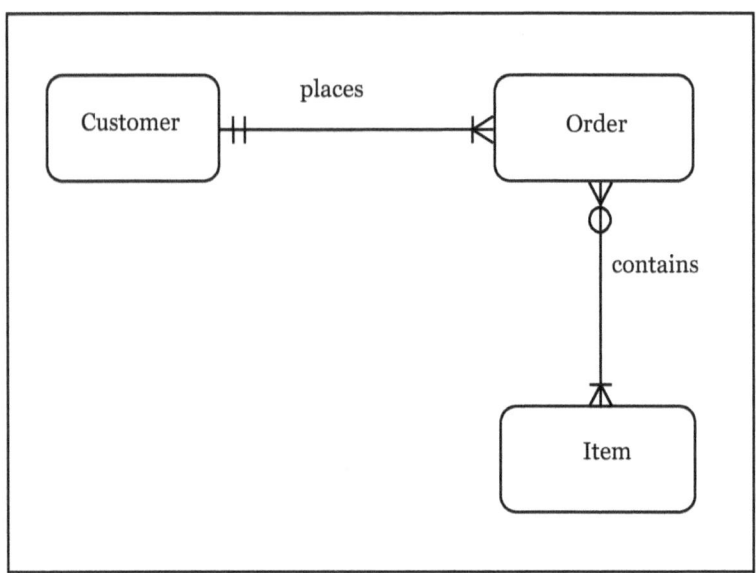

Event List

1. Customer Places Order
2. Customer Requests Item Availability
3. Customer Requests Order Status
4. *Time to Generate Weekly Sales Report*
5. Management Submits Item Discount
6. Customer Returns Item

System Response

1. Fill Customer Order
2. Return Item Availability
3. Return Order Status
4. ***Generate Weekly Sales Report***
5. Apply Item Discount
6. Return Item

Lean Use Case

Generate Weekly Sales Report is initiated when a specified time is reached. The system must respond by reporting all sales for the previous week.

Use Case Diagram

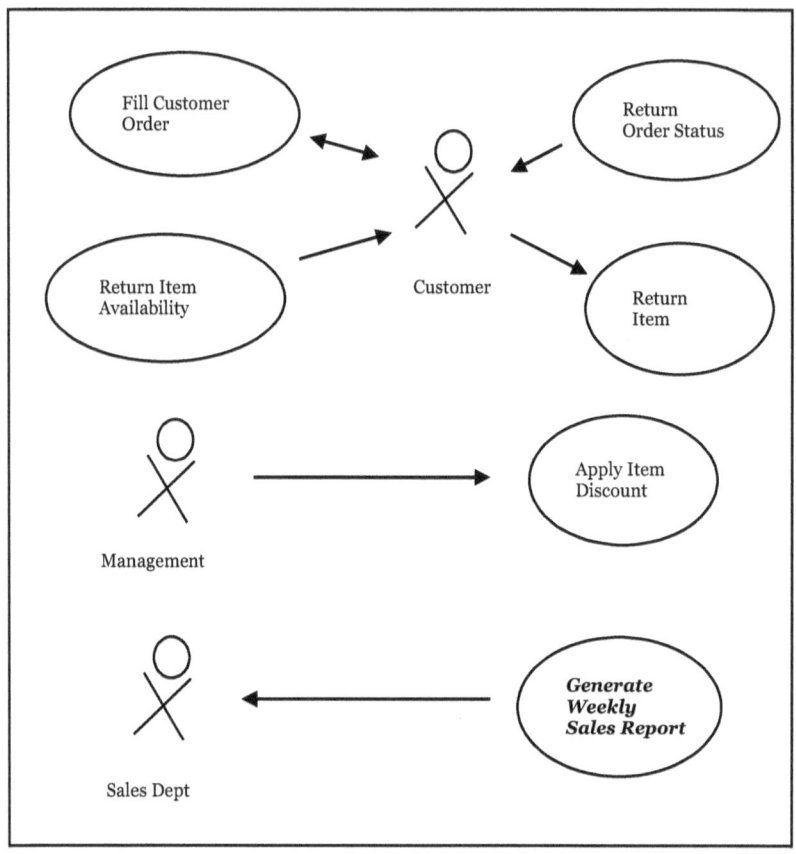

Use Case JIT Detail

1. *Generate Weekly Sales Report*

2. UC # 2; Wiley; created November 2009; Web Order System (WOS)

3. Priority: 5

4. Goal: to provide a report of sales for the previous week.

5. *Generate Weekly Sales Report* is initiated when a specified time is reached. The system must respond by:

    1) reporting all sales for the previous week,

2) deriving order status for each new order.

6. Use Case Trigger (business event): *Time to Generate Weekly Sales Report* (a point in time is reached)

7. Actors: Sales Department

8. Preconditions: none.

9. Primary Scenario:

    1) time to generate weekly sales report

    2) system derives dates for reporting week

    3) system pulls orders from reporting week

    4) system associates customer and items with each order to be reported

    5) system derives status of each order to be reported

    6) system generates report

    7) system ends reporting session

10. Alternative Paths

    Alternative path #1: no orders for reporting week

    1) leave primary path following step 3

    2) system reports no orders for reporting week

    3) system ends reporting session

11. Post Conditions:

    1) report received by Sales Department

    2) data stores are unchanged

12. Business Rules:

    1) eligible orders will have a submission date that falls on or within reporting week

13. Input Summary:

    1) weekly date range (system derived)

14. Output Summary:

    1) order number

2) order submission date

3) order status

4) order purchase total

5) number and description of each item on order

6) customer data

15. Use Case Issues/Risks:

1) none

16. Use Case Notes:

1) none

17. Non-Functional Requirements:

1) security: none

2) performance: none

Requirements Verification

Changes at this point are many times less costly to make than those detected later in the lifecycle. Once the Use Case definition is complete, the requirements for **Generate Weekly Sales Report** can be verified and the UAT can be designed.

Design, Build, Test

Focus for the design, build, and test effort can be on the single module **Generate Weekly Sales Report** reducing the overall complexity of these tasks.

User Acceptance Test

Once the Use Case is verified and the UAT written, the **Generate Weekly Sales Report** software can be user tested and either repaired or prepared for deployment.

System Test, Deployment

Following the priority established in the Detail Event Use Case, ***Generate Weekly Sales Report*** can be system tested and readied for deployment.

Chapter 20
Apply Item Discount

Apply Item Discount is initiated by management to support a campaign to discount the price of specified items. The system must respond by applying new pricing data for a specified item.

Context Diagram

The Event Thread for **Apply Item Discount** has its beginning in this model as management submits discount data for an item.

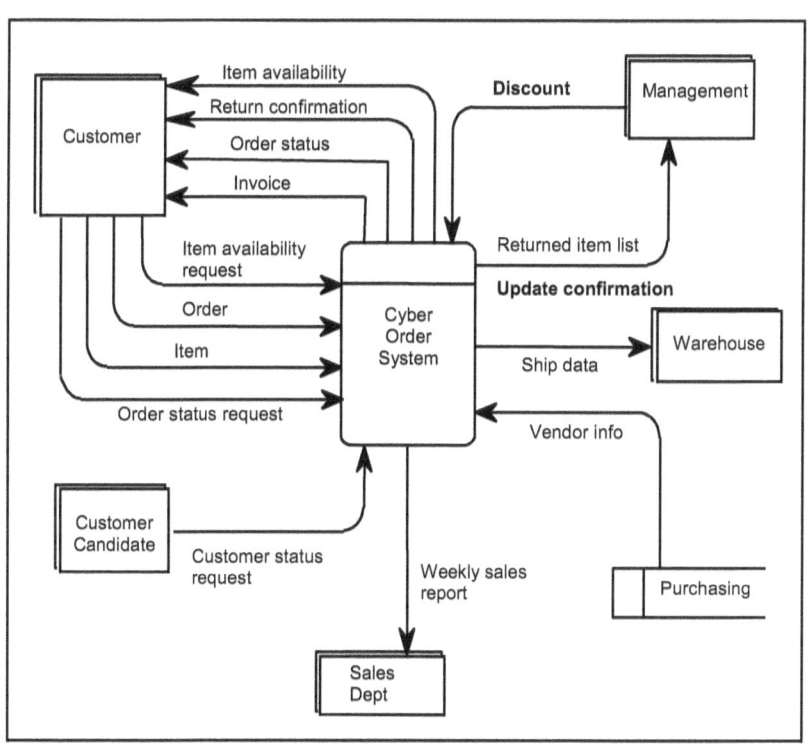

Lean Data Model

As entities and relationships are identified, more potential business events are revealed. The data model fragment must support application of discount data by date associated to item. This will require the beginning of normalization of the data model as an attributive ('weak' or 'dependent') entity will be added for discount data (refer to Appendix D).

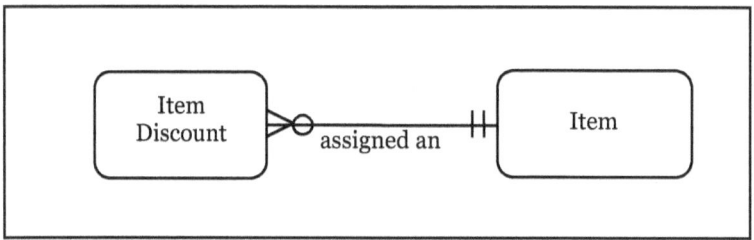

Event List

1. Customer Places Order
2. Customer Requests Item Availability
3. Customer Requests Order Status
4. Time to Generate Weekly Sales Report
5. ***Management Submits Item Discount***
6. Customer Returns Item

System Response

1. Fill Customer Order
2. Return Item Availability
3. Return Order Status

4. Generate Weekly Sales Report
5. *Apply Item Discount*
6. Return Item

Lean Use Case

Apply Item Discount is initiated by management to support a campaign to discount the price of specified items by lowering the price of specified items.

Use Case Diagram

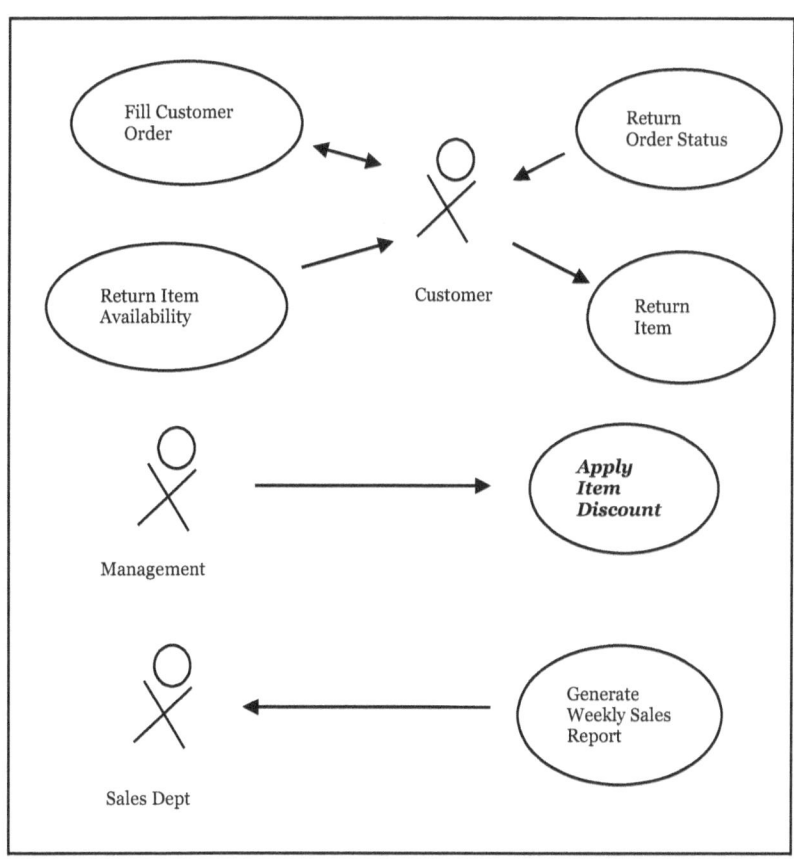

Use Case JIT Detail

1. *Apply Item Discount*

2. UC # 2; Wiley; created November 2009; Web Order System (WOS)

3. Priority: 6

4. Goal: to apply new pricing of specified items.

5. *Apply Item Discount* is initiated by management to support discounting of specified items. The system must respond by:

    1) verifying management,

    2) verifying item number,

    3) rejecting an invalid item number with an option to correct,

    4) verifying pricing data,

    5) rejecting invalid pricing data with an option to correct,

    6) applying new pricing data for the specified item.

6. Use Case Trigger (business event): *Management Submits Item Discount* (discount pricing)

7. Actors: management

8. Preconditions:

    1) items to discount must be on file

    2) management electronic signature must be on file

9. Primary Scenario:

    1) management selects 'apply item discount data'

    2) system verifies management

    3) management submits discount data

    4) system verifies item number

    5) system verifies pricing data

    6) system applies new pricing data

    7) system reports pricing updates

       8) management ends query session

10. Alternative Paths

    Alternative path #1: Invalid item

       1) leave primary path following step 3

       2) system reports invalid item

       3) system offers opportunity to correct item number

       4) management submits corrected item number

       5) return to primary path step 3

    Alternative path #2: Invalid discount data

       1) leave primary path following step 4

       2) system reports invalid pricing data

       3) system offers opportunity to correct pricing data

       4) management submits corrected pricing data

       5) return to primary path step 4

11. Post Conditions:

    1) pricing data updated

    2) management has received discount data update report

12. Business Rules:

    1) only management can update pricing data

    2) after two failed tries, session will be dropped

    3) discount dates cannot overlap for a specific item

13. Input Summary:

    1) item number

    2) discount data

    3) valid discount date range

    4) management signature

14. Output Summary:

    1) item number

    2) item description

3) old pricing data

4) new pricing data

5) date of update

6) signature of management submitting pricing changes

15. Use Case Issues/Risks:

1) how many invalid item numbers to allow before dropping session

2) how many invalid pricing updates to allow before dropping session

16. Use Case Notes:

1) none

17. Non-Functional Requirements:

1) security: management electronic signature must be on file

2) performance: response to user will be less than 5 seconds

Requirements Verification

Changes at this point are many times less costly to make than those detected later in the lifecycle. Once the Use Case definition is complete, the requirements for **Apply Item Discount** can be verified and the UAT can be designed.

Design, Build, Test

Focus for the design, build, and test effort can be on the single module **Apply Item Discount** reducing the overall complexity of these tasks.

User Acceptance Test

Once the Use Case is verified and the UAT written, the **_Apply Item Discount_** software can be user tested and either repaired or prepared for deployment.

System Test, Deployment

Following the priority established in the Detail Event Use Case, **_Apply Item Discount_** can be system tested and readied for deployment.

Chapter 21
Return Item

Return Item is initiated by the customer when an item is returned. The system must respond by applying the return data.

Context Diagram

The Event Thread for **Return Item** has its beginning in this model as the customer returns an item.

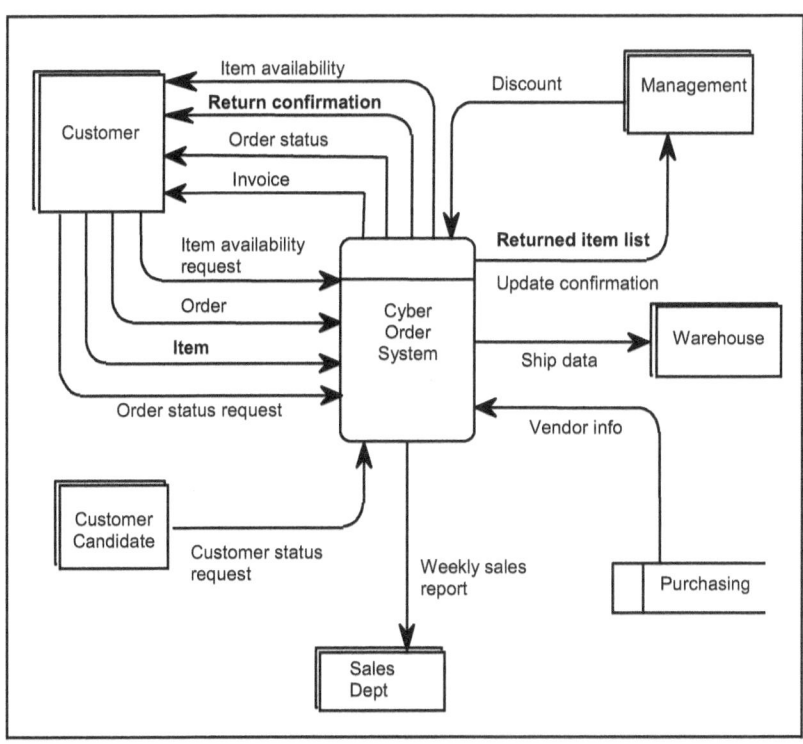

Lean Data Model

As entities and relationships are identified, more potential business events are revealed. The data model fragment must support return data associated with an ordered item. This will require normalization as an attributive entity to hold item return data is required (see Appendix D).

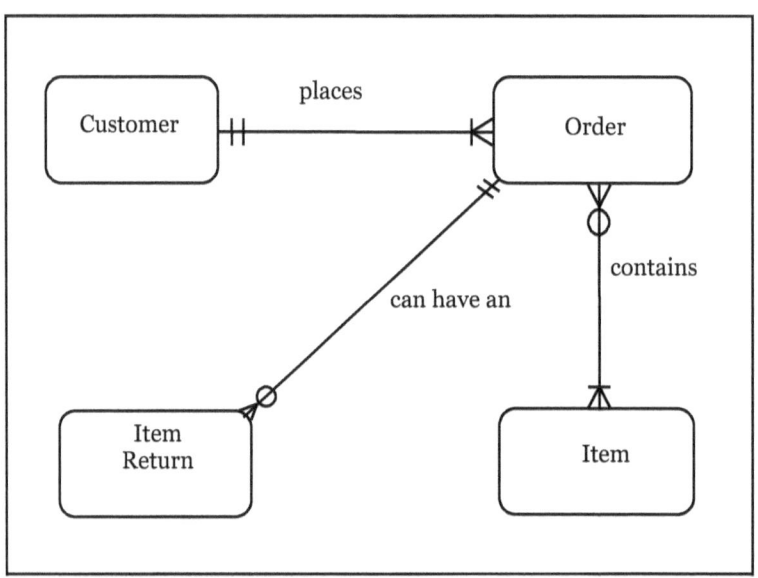

Event List

1. Customer Places Order
2. Customer Requests Item Availability
3. Customer Requests Order Status
4. Time to Generate Weekly Sales Report
5. Management Submits Item Discount
6. ***Customer Returns Item***

System Response

1. Fill Customer Order
2. Return Item Availability
3. Return Order Status
4. Generate Weekly Sales Report
5. Apply Item Discount
6. ***Return Item***

Lean Use Case

Return Item is initiated by the customer when an item is returned. The system must respond by confirming the validity of the return and updating data stores.

Use Case Diagram

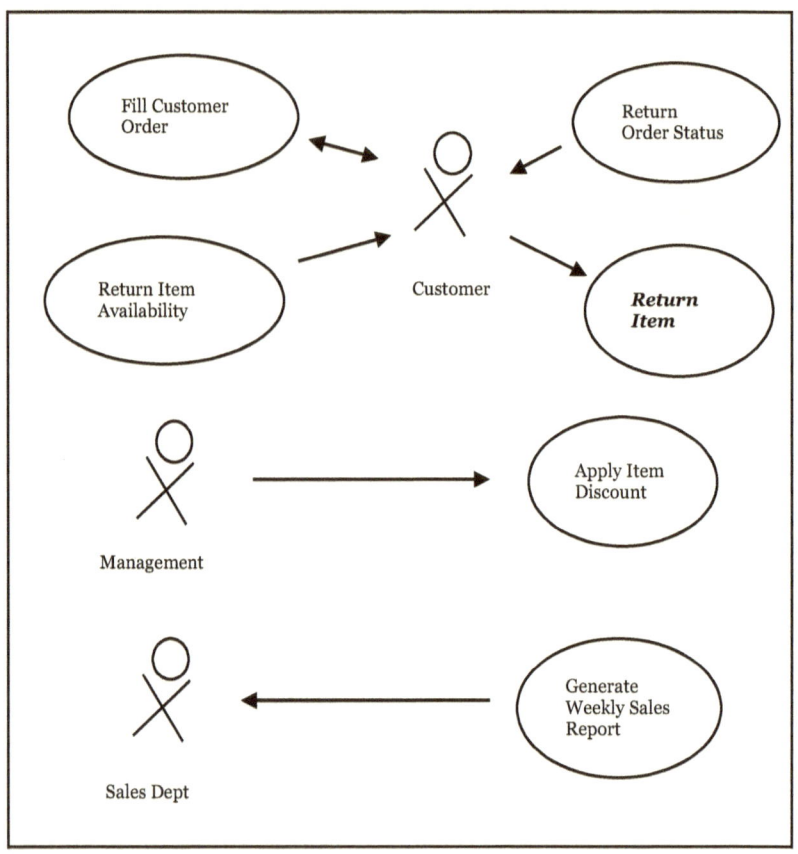

Use Case JIT Detail

1. ***Return Item***

2. UC # 2; Wiley; created November 2009; Web Order System (WOS)

3. Priority: 2

4. Goal: to return an item to inventory and report the return.

5. *Return Item* is initiated by the customer by returning a purchased item. The system must respond by:

    1) verifying the item number,

    2) rejecting an invalid item number with an option to correct,

    3) verifying the order and return data,

    4) reporting the return,

    5) updating appropriate data stores,

    6) confirming item return.

6. Use Case Trigger (business event): *Customer Returns Item* (returned item)

7. Actors: customer, management

8. Preconditions:

    1) Customer and item must be on file.

    2) Returned item must be associated with a valid order.

9. Primary Scenario:

    1) customer selects 'return item'

    2) customer submits item to return

    3) system checks customer status

    4) system verifies item number

    5) system verifies associated order

    6) system updates quantity-on-hand

    7) system updates returned item data store

    8) system generates line-item on returned item report

    9) customer ends session

10. Alternative Paths

    Alternative path #1: Customer not on file

        1) leave primary path following step 3

        2) system reports customer not on file

3) system offers opportunity to establish customer profile

4) customer submits customer profile data

5) system validates customer profile

6) return to primary path step 4

Alternative path #2: Invalid item

1) leave primary path following step 4

2) system reports invalid item

3) system offers opportunity to correct item number

4) customer submits corrected item number

5) return to primary path step 4

Alternative path #3: Invalid order

1) leave primary path following step 5

2) system reports invalid order

3) system offers opportunity to correct order number

4) customer submits corrected order number

5) return to primary path step 5

11. Post Conditions:

1) returned item data stores are updated

2) returned item report update is generated

12. Business Rules:

1) only customers on file can return an item

2) item to be returned must be associated with a valid order

3) after two failed tries, session will be dropped

13. Input Summary:

1) customer id

2) item number

3) order number

14. Output Summary:

1) updated quantity-on-hand
2) returned item description
3) returned item quantity
4) reason for return
5) returned item report

15. Use Case Issues:
    1) how many invalid item numbers to allow before dropping session
    2) how many invalid customer ids to allow before dropping session
    3) how many invalid order numbers to allow before dropping session

16. Use Case Notes:
    1) none

17. Non-Functional Requirements:
    1) security: customer must be on file
    2) item must be on file
    3) order must be on file
    4) performance: response to user will be less than 5 seconds

Requirements Verification

Changes at this point are many times less costly to make than those detected later in the lifecycle. Once the Use Case definition is complete, the requirements for **Return Item** can be verified and the UAT can be designed.

Design, Build, Test

Focus for the design, build, and test effort can be on the single module **Return Item** reducing the overall complexity of these tasks.

User Acceptance Test

Once the Use Case is verified and the UAT written, the **Return Item** software can be user tested and either repaired or prepared for deployment.

System Test, Deployment

Following the priority established in the Detail Event Use Case, **Return Item** can be system tested and readied for deployment.

Chapter 22
Test Cases

User acceptance tests (UAT) are the only tests designed to verify requirements. A UAT test case is developed for a single Event Use Case scenario; each Use Case scenario must be tested.

One objective of testing is to exercise every path (scenario) through each Use Case. Test cases contain at least the components shown below.

- o Use Case steps
- o test procedure steps
- o input data
- o database data
- o expected results
- o test notes

Four test cases taken from the running case study Use Case *Fill Customer Order* demonstrate the level of detail required for adequate testing.

- o Order Successful (Scenario 1)
- o Insufficient Stock, Backorder Rejected (Scenario 2)
- o Insufficient Stock, Backorder Accepted (Scenario 3)
- o Invalid Order (Scenario 4)

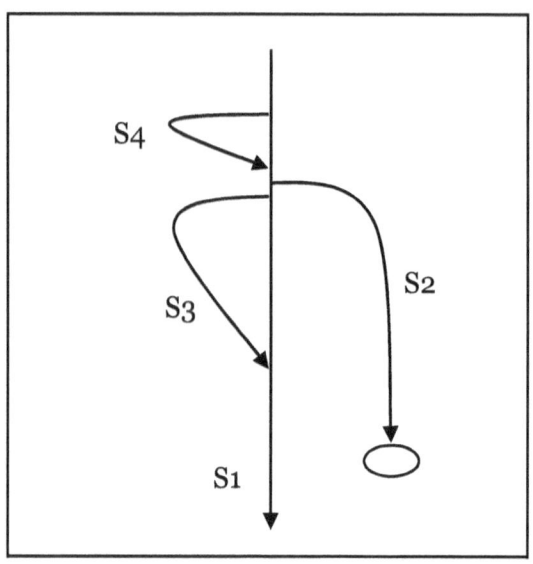

There are many others, for example, out-of-stock where in-stock quantity is zero. View of test database contents must be available for test execution.

Order Successful

Scenario 1 (Use Case steps)

1. select 'place order' from menu
2. customer submits order information
3. system validates order information
4. system checks for sufficient quantity of product
5. system requests pay information from customer
6. customer submits pay information
7. system exports pay data
8. system updates data stores
9. system generates pick list, invoice, shipping documents

10. system informs customer of successful completion of order

11. customer closes order session

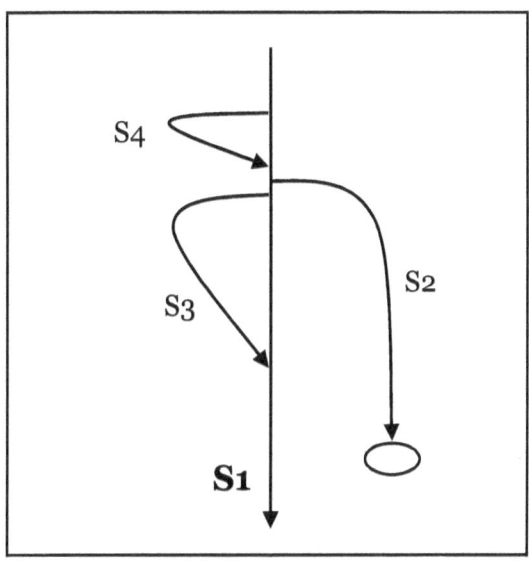

Scenario 1, Order Successful, Test Case 1

Procedure Step	Input Data	Expected Results	Test Notes
1) select from menu to place order	'place order' choice	- product order screen is presented	
2) submit order data	- customer ID = valid ID* - product number = valid number* - order quantity = n where n is less than the database value	- data valid message - sufficient quantity message - pay information request	
3) provide pay information	- valid card type* - card number* - card expiration date*	- card valid message - order successful message - system generates pick list, invoice, shipping documents - system updates data stores - place new order option offered	
4) end order session	end session screen selection	- system ends normally	

* choose from test database

Insufficient Stock/Backorder Rejected

Scenario 2 (Use Case steps)

1. leaves main flow (S1) following step 4
2. system informs customer of insufficient product quantity
3. system offers customer product backorder choice
4. customer rejects product backorder option
5. system offers option to order another product
6. customer ends order session

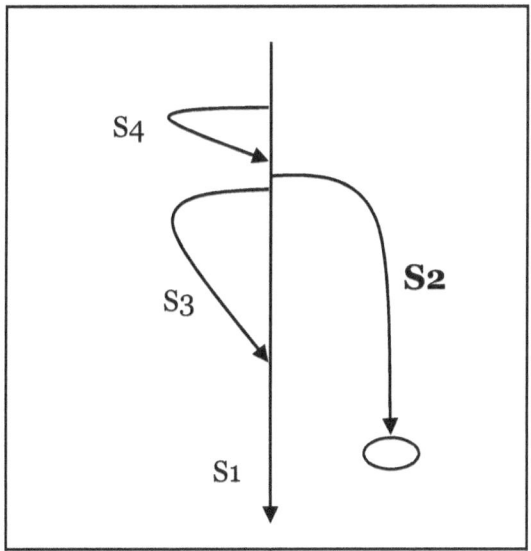

Scenario 2: Insufficient Stock/Backorder Rejected
Test Case 1

Procedure Step	Input Data	Expected Results	Test Notes
1) select from menu to place order	'place order' choice	- product order screen is presented	
2) submit order data	- customer ID = valid ID* - product number = valid number* - order quantity = n where n is greater than value in database	- data valid message - insufficient quantity message - backorder option offered	
3) select no backorder	'no backorder' choice	- backorder rejected - place new order option offered	
4) end order session	end session screen selection	- system ends normally	

* choose from test database

Insufficient Stock/Backorder Accepted

Scenario 3 (Use Case steps)

1. leaves Scenario 1 following step 4
2. system informs customer of insufficient product quantity
3. system offers customer product backorder option
4. customer accepts product backorder option
5. return to Scenario 1 step 5

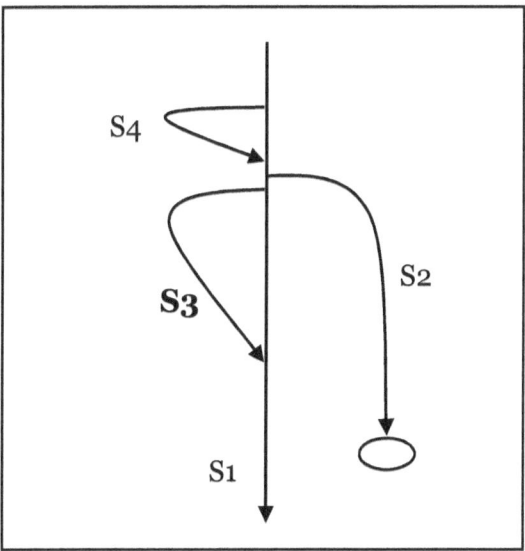

Scenario 3: Insufficient Stock/ Backorder Accepted

Test Case 1

Procedure Step	Input Data	Expected Results	Test Notes
1) select 'place order' from menu	'place order' choice	- product order screen is presented	
2) submit order data	- customer ID = valid ID* - product number = valid number* - order quantity = n where n is greater than value in database	- data valid message - insufficient quantity message - backorder option offered	
3) select backorder	'backorder' choice	- backorder accepted - pay information request	
4) provide pay information	- valid card type* - card number* - card expiration date*	- card valid message - backorder successful message - place new order option offered	
5) end order session	end session screen selection	- system ends normally	

* choose from test database

Invalid Order

Scenario 4 (Use Case steps)

1. leaves Scenario 1 following step 3
2. system reports invalid order
3. system offers corrective action
4. customer submits corrected order
5. return to Scenario 1 step 3

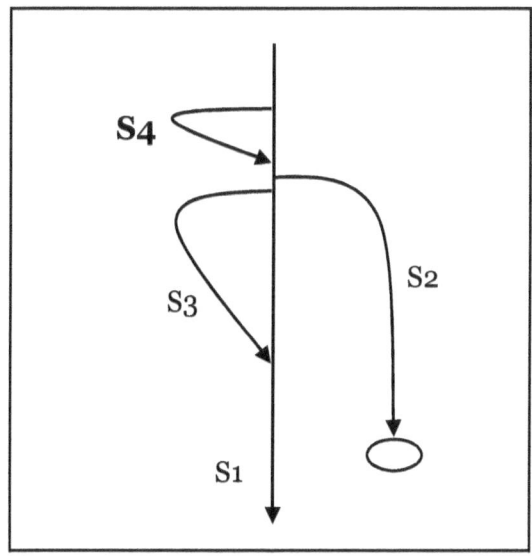

Scenario 4: Invalid Order, Test Case 1

Procedure Step	Input Data	Expected Results	Test Notes
1) select 'place order' from menu	'place order' choice	- product order screen is presented	
2) submit order data	- customer ID = valid ID* - product number = invalid number* - order quantity = n where n is less than database value	- invalid order message - corrective action offered	
3) submit corrected data	- customer ID = valid ID* - product number = valid number* - order quantity = n where n is less than database value	- data valid message - pay information request	
4) provide pay information	- valid card type* - card number* - card expiration date*	- card valid message - order successful message - place new order option offered	
5) end order session	end session screen selection	- system ends normally	

* choose from test database

Book 3 Epilogue

Book 3 has presented the artifacts of a running case study. Six events have been developed to demonstrate how the Thread partitions the build phase. Test examples show the level of detail required for a thorough test effort.

Beginning with the conceptual phase where the events initially emerge, the Event Threads have partitioned the effort and focused the teams on singular efforts of manageable size. These Threads remain recognizable from the earliest tasks to implementation and maintenance. They tame the complexity common in an information system.

Summary of Key Concepts

While not a new concept, events are one of the most effective (and overlooked) ways to define and manage requirements. But more important than the events are the "Threads" that result. With each event, an Event Thread is born.

These Threads bind the early and late stage artifacts and partition the entire system development effort both horizontally (across system scope) and vertically (across the lifecycle). Effectively, each Thread has its own lifecycle as the artifacts of one event are relatively isolated from other event partitions by the independent Threads.

The conceptual and physical layers are brought to a single focus (a Thread) as the development progresses through analysis, design, build, test, and deployment. An event defines a relatively small piece of the problem while its Thread establishes the lifecycle for that piece.

The following are key concepts that the reader can learn from this book.
- o Events partition naturally;
- o Events are middle-out not top down;
- o Events have a strong user connection;
- o Each event is the beginning of a Thread;
- o Each Thread has its own lifecycle and artifacts;

- An Event Thread preserves the identity and traceability of the event throughout the lifecycle;
- The response to an event is planned;
- An event and the resulting Thread define a Use Case;
- Just-in-time Use Cases provide for an agile approach.

Appendices

A.

Methodology Conceptual Phases

B.

Methodology Artifacts – Single Event Thread

C.

The Thread and the Lifecycle

D.

Normalized Data Model

Appendix A
Methodology Conceptual Phases

The conceptual phase of the methodology is comprised of an early waterfall phase (Appendix A1) and an iterative, just-in-time (JIT) phase (Appendix A2). The early waterfall phase takes a one-pass, high-level look at the proposed system while the later phase focuses on a single Use Case per iteration. Fresh requirements are passed downstream to the design, build, test, and deploy steps one Use Case at a time.

The following acronyms are used in the Appendix A diagrams.

CRUD: Create, Read, Update, and Delete (CRUD) are the four basic functions of persistent storage. Sometimes CRUD is extended to CRUDA to include archive.

EUC: Event Use Case; a Use Case based on and whose scope is defined by an event.

JIT: Just In Time; follows a manufacturing strategy to minimize inventory in which it arrives just as needed.

SRT: System Response Table; a tabular listing of the system's responses to events.

Broken line: ------------ a diagrammatic representation of a physical task or physical data flow.

Appendix A1
Early Conceptual Phase (Waterfall)

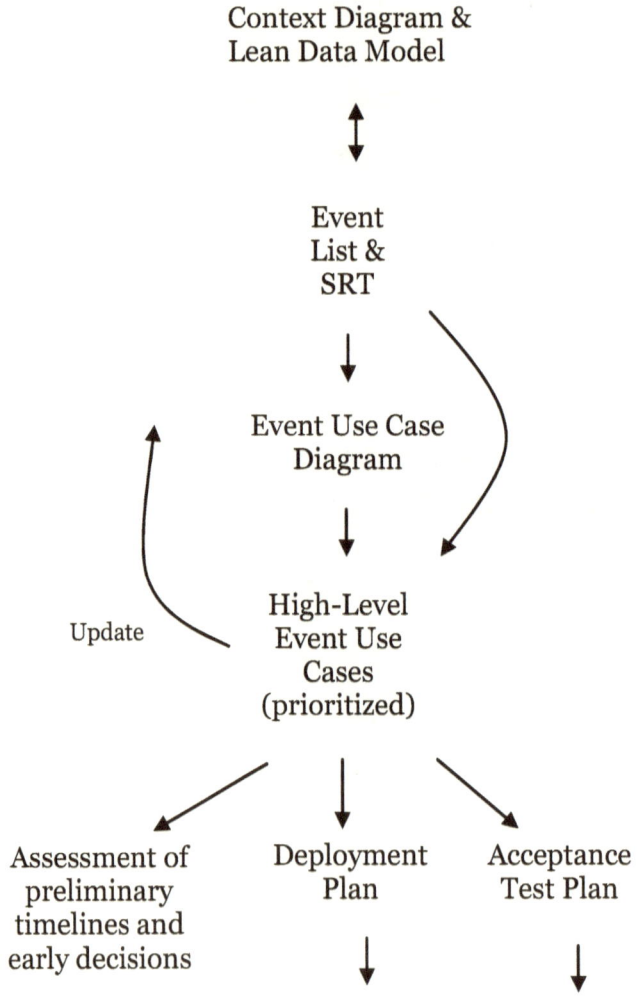

Context Diagram &
Lean Data Model

Event
List &
SRT

Event Use Case
Diagram

Update

High-Level
Event Use
Cases
(prioritized)

Assessment of
preliminary
timelines and
early decisions

Deployment
Plan

Acceptance
Test Plan

Appendix A2
Late Conceptual Phase (EUC JIT Iterations)

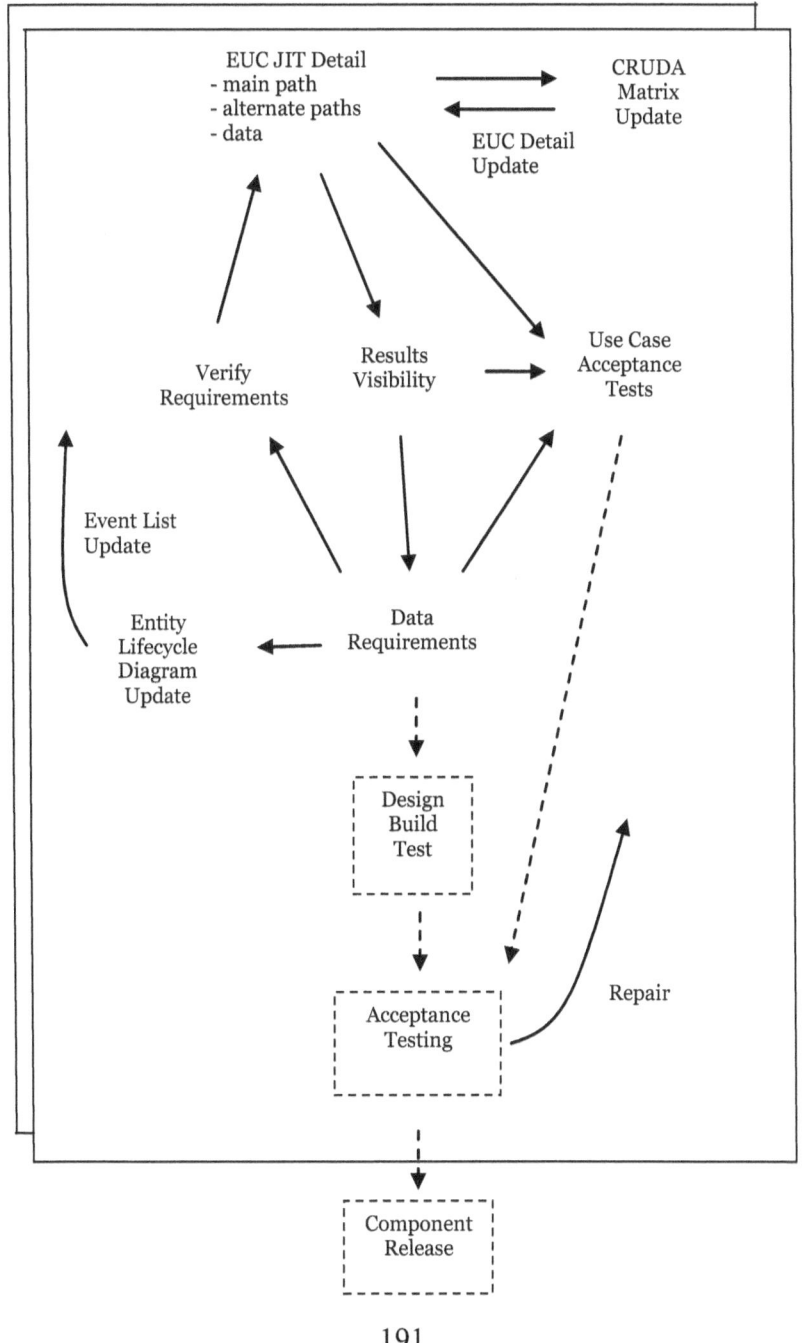

Appendix B
Methodology Artifacts - Single Event Thread

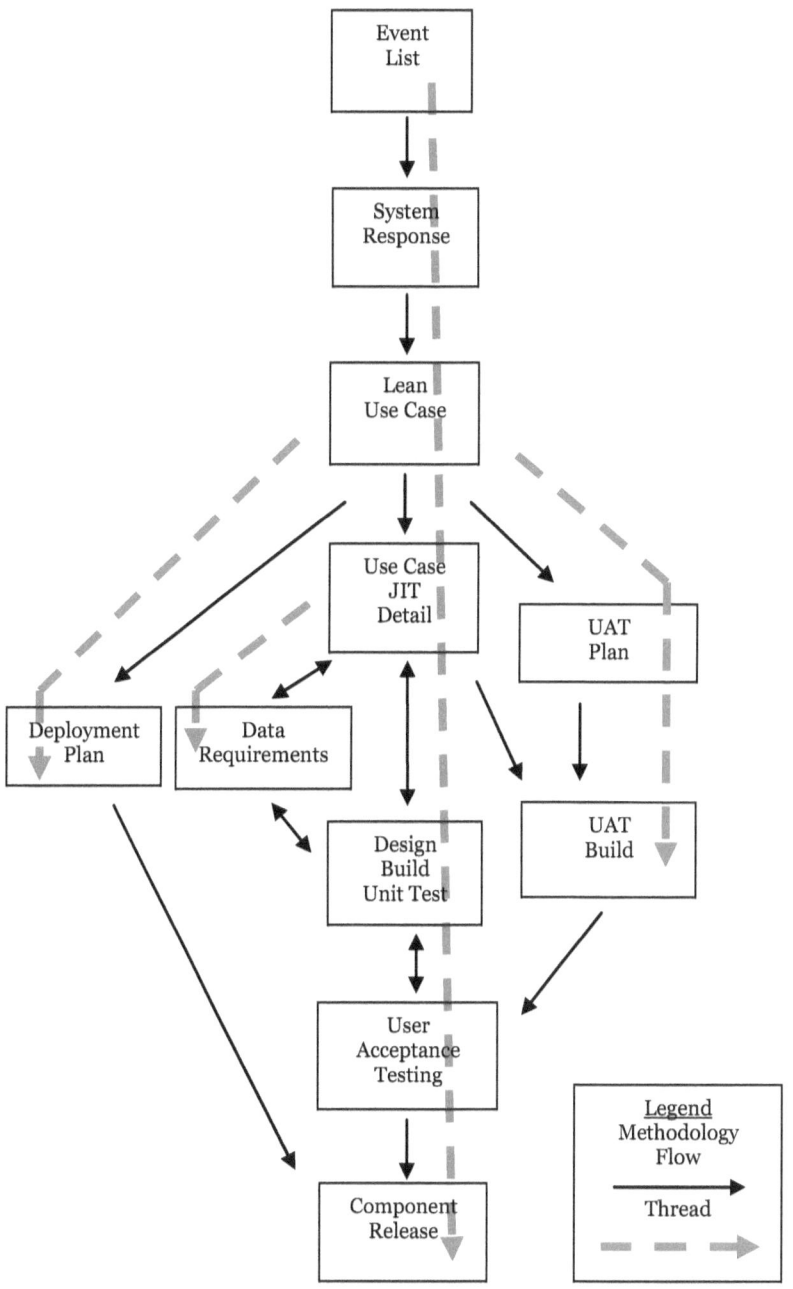

Appendix C
The Thread and the Lifecycle

A Thread extends vertically through the lifecycle from requirements definition to the deployment of system components and production. An Event Thread slices through the entire development process. *Customer Places Order* and *Customer Requests Order Status* are depicted below as each phase of the lifecycle addresses a single event.

	Customer Places Order	Customer Requests Item Availability	Customer Requests Order Status	Time to Generate Weekly Sales Report	Management Submits Item Discount Data	Customer Returns Item
Requirements						
Design						
Construction						
Testing						
Deployment						
Maintenance						

Appendix D
Normalized Data Model

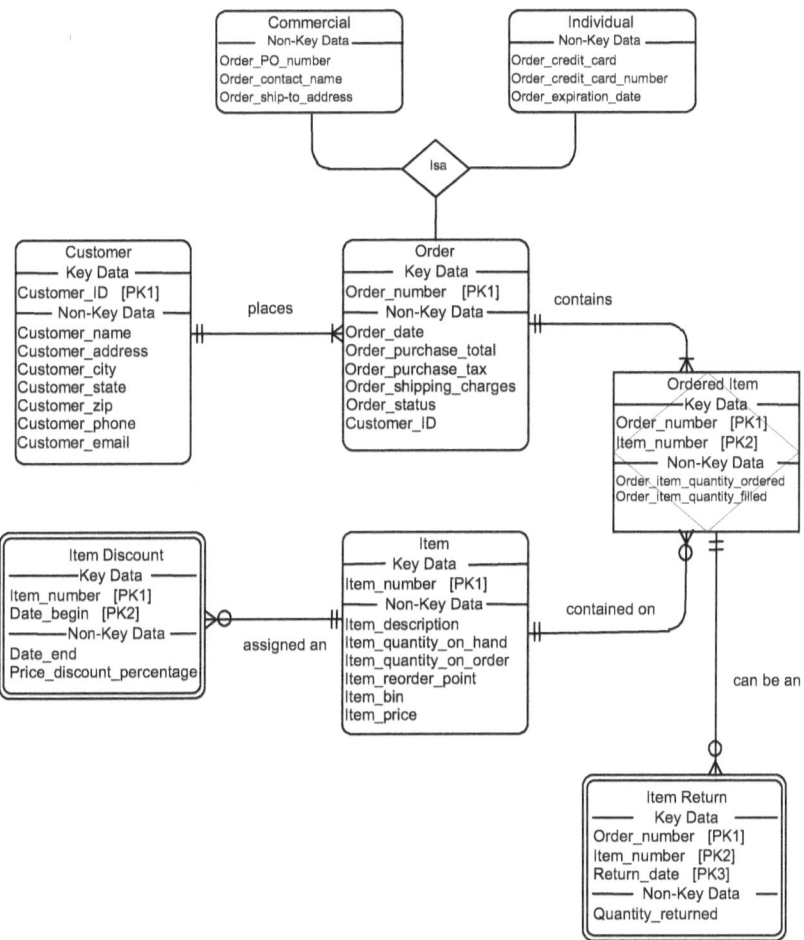

Glossary

actor: A term meaning user or agent of the system.

Associative Entity: A data entity that resolves a many-to-many relationship by defining the association between two related entities.

Attributive Entity: A data entity that holds attributes that describe another entity. These attributes may repeat or may be null for a single instance of the original entity key. (synonyms: weak entity, dependent entity)

cohesion: The degree to which a system component accomplishes one and only one function.

consistent state: Data consistency states that data cannot be written to a database that would violate the database's own rules for valid data. This includes entity, referential, and domain integrity.

coupling: The degree to which one system component is dependent upon another.

CRUDA: A matrix that depicts the interaction between the data entities from the data model and the event responses by recording which of the custodial functions (create, read, update, delete, and archive) are applied to the data by the processes.

data model fragment: Entities and relationships that are relevant to a specific part of the overall solution (one Use Case) and will eventually contribute to the total data model.

ELCD: Entity Lifecycle Diagram; represents entity states, entity dependencies, and entity state changes.

event: An activity in the user's environment that requires a response from the proposed information system.

event trigger: A data flow that occurs in response to a business event and flows into the system, invoking the planned event response.

EUC: An Event Use Case; a Use Case that is derived from an event response and whose scope is determined by the required event response processing.

Extend Use Case: A Use Case that contains processing common to more than one Use Case and is executed from those Use Cases; this code exists only one time thus avoiding duplicate code.

JIT: Just-in-time; an approach that produces a product just before it is needed.

persistent: Existing for a long or longer than usual time.

pervasive: Existing in or spreading through every part of something.

SME: Subject Matter Expert.

SRT: System Response table.

results visibility: Hard copy, screen based reporting.

UAT: User Acceptance Test; a test used to verify that the system was built with the correct requirements.

vertical extension of a Thread: Extending from early in the lifecycle to deployment and production phases.

References

DeSmedt, William, "The Wolf at the Door," *Database Programming & Design*, April 1994, p64.

McMenamin, Stephen, and John Palmer, *Essential Systems Analysis*, Englewood Cliffs, N.J.: Prentice Hall, 1984.

Ramo, Joshua Cooper, *The Age of the Unthinkable*, New York, New York: Little, Brown and Company, 2009.

Rumbaugh, James, Michael Blaha, William Premerlani, Frederick Eddy, and William Lorenson, *Object-Oriented Modeling and Design*, Englewood Cliffs, N.J.: Prentice hall, 1991.

Wiley, Bill, *Essential System Requirements*, Reading, Massachusetts: Addison-Wesley, 2000.

FSTda627600, Use Case and Scenario Design Research, Use Case, https://sites.google.com/site/fstda627600/researchmax/use-case-and-scenario-design-research, Date accessed: August, 2014.

Index

207

Notes

Notes

Notes

www.ingramcontent.com/pod-product-compliance
Lightning Source LLC
Chambersburg PA
CBHW030930180526
45163CB00002B/521